Do you have
the knowledge?

Can you take
a position?

Will you lead
the debate?

Mobile Policy Handbook

An insider's guide
to the issues

About this Handbook

A country's citizens benefit most when the private and public sectors work together in a spirit of openness and trust. To this end, the GSMA is committed to supporting governments and regulators in their efforts to introduce pro-investment telecommunications policies.

The Mobile Policy Handbook: An Insider's Guide to the Issues is part of the GSMA's efforts to promote such collaboration. A unique resource that assembles a range of policy topics and mobile industry positions and initiatives under one cover, it acts as a signpost to regulatory best practice.

As the global trade association of mobile operators, the GSMA conducts and commissions research on policy trends and challenges in the mobile communications market. This handbook draws on the association's unique insight into the mobile sector and presents it in a practical way for those who want to explore the issues and unleash the value of mobile technology in their own market.

In this fourth edition of the Mobile Policy Handbook, the Mobile for Development section has been expanded and a number of new policy topics and industry positions have been added. The Mobile Initiatives section has also been reorganised and refreshed to reflect the broad priority areas for the industry.

The online version of this resource — **www.gsma.com/publicpolicy/handbook** — offers an always up-to-date catalogue of the mobile industry's policy positions. Readers are encouraged to contact the GSMA if they have any questions or requests for more information. E-mail us at **handbook@gsma.com.**

World-Changing Trends

Two technologies have transformed the lives of billions of people over the past two decades — mobile communications and the internet. Initially, these technologies developed in parallel, but now they are on a fully converged path.

This convergence heralds a new era, with the majority of the world's population not only making their first phone call using a mobile handset, but accessing the internet over mobile technology too. Equally profound is the revolution in machine-to-machine communications. We are at the very beginning of this development, but already billions of automated messages flow between widely connected devices, over the internet, driving forward productivity and making major improvements in health services, for example. Today over half of the world's population has access to a mobile phone; by the end of this decade, the mobile internet will support over one billion machine-to-machine connections.

These dominant trends drive much of the GSMA's work with policymakers, bringing into new focus issues such as data protection and privacy, the internet of things, network economics and mobile government.

Never before has the role of the communications ministry and regulator been so critical to the success of governments' economic and social policies — with implications for business, education, health, access to financial and government services, and so much more.

As the mobile internet becomes the key to the transformation of many other sectors, policymakers face new and exciting challenges and will need to navigate uncharted waters. We hope this handbook provides a compass that is referred to regularly on that voyage.

Mobile for Development

The astonishing growth of mobile puts the industry in a unique position to enhance the lives of underserved people in emerging markets. GSMA Mobile for Development (M4D) places mobile at the centre of innovation in commercial services to help deliver sustainable solutions that have strong socio-economic impact on the lives of these communities.

M4D targets issues such as financial and digital exclusion, the lack of official identity provision, inadequate availability of healthcare, the gender gap, limited access to energy and water, low agricultural yields and the need for effective disaster response during times of crisis.

For example, M4D is helping drive financial inclusion by working to identify ways in which mobile money services can achieve scale and deliver convenient and affordable financial services to the underserved. Similarly, by supporting the efforts of mobile operators to expand their networks in rural areas it is helping bridge the digital divide and through its work on mobile identity solutions it is helping to close the identity gap for huge numbers of people in developing nations who currently lack official identity.

M4D is also supporting the development of smarter infrastructure in emerging markets, for example, by leveraging embedded SIM technology in water pumps, and is working with agricultural organisations and the development community on the creation of scalable and commercially sustainable agricultural mobile services. These efforts and more are detailed over the following pages.

Connected Society

Background

Approximately 62 per cent of the world's population is not connected to the internet, and the majority of these people are rural consumers in developing countries. This leaves them excluded from social and economic opportunities, and negatively impacts the economic development of the countries in which they live.

For example, management consulting firm McKinsey estimates that the internet could account for as much as ten per cent of Africa's annual gross domestic product by 2025 (up from around one per cent today), due to the internet's transformational effects on retail, agriculture, education and healthcare.

In the developing world, mobile is the cheapest and most convenient way of accessing the internet, and is often the first internet touch point for rural consumers. According to GSMA Intelligence, mobile penetration in the developing world has increased steadily from approximately 25 per cent to 44 per cent over the past five years.

However, despite this rapid increase, a number of barriers still stand in the way of these populations fully benefiting from the mobile internet. The barriers include infrastructure challenges, affordability, and consumer impediments such as digital literacy and the lack of locally relevant content

Programme Goals

The GSMA's Connected Society programme supports the mobile industry in its efforts to bring mobile internet access to underserved populations in developing countries. Working towards this aim, the programme delivers evidence-based research, case studies and advisory services to help mobile operators, policymakers and other stakeholders break down barriers related to infrastructure, affordability, digital literacy and local content.

The programme also works closely with the GSMA Connected Women initiative to close the gender gap in mobile phone ownership. Furthermore, it collaborates with the GSMA Disaster Response programme in its goal of helping humanitarian organisations, governments, non-governmental organisations (NGOs) and the broader mobile ecosystem develop coordinated and highly effective disaster response mechanisms for the mobile industry.

Public Policy Considerations

The growth of the mobile sector has enabled broader access to Information and Communications Technology (ICT), enabling information sharing, increased productivity and ultimately economic and social growth. However, barriers to accessing the mobile internet and services remain, especially in developing countries. There are a number of areas that governments and other key stakeholders can focus on to help bridge this digital divide.

Affordability. Mobile-specific taxes raise barriers to mobile phone ownership and usage. This is especially true in developing markets where affordability is critical to increasing digital inclusion and access to the mobile internet. Reducing taxation on mobile access lowers these barriers. By adopting phased reductions in these taxes, governments can benefit from the additional economic growth driven by the mobile internet, while limiting short-term losses in tax revenue.

Digital literacy. The future growth in mobile internet adoption will come from infrequent or new internet users. However, many new internet users have limited digital skills preventing them from gaining the full benefit of the mobile internet. Educational efforts by key stakeholders (including operators, NGOs and governments) can help address this issue.

Infrastructure. Currently, 2G mobile networks cover 95 per cent of the world's population, and 3G coverage has reached 69 per cent. However, 3G coverage is much less extensive in rural areas, at just 29 per cent, reflecting the greater practical and economic challenges related to providing power and backhaul links in remote locations, as well as difficulties in carrying out network maintenance in these areas. To bridge the digital divide the overall regulatory environment must encourage investment in mobile infrastructure and the use of mobile services. One of the biggest barriers is the availability of affordable spectrum. Governments can help here by releasing sufficient spectrum at affordable cost, especially lower-frequency spectrum, as it allows operators to serve larger areas with fewer base stations. Support for industry-led solutions to infrastructure sharing can also help improve the economics of rural coverage.

Locally relevant content. In many developing nations there is a lack of locally relevant content to attract people to regularly use the mobile internet. Governments have an important role to play in encouraging key enabling infrastructure such as affordable international bandwidth and local content hosting platforms. Operators, governments, NGOs, development organisations and donors should also work together to break down barriers to the creation of relevant material by equipping local people with the skills needed to produce content. Governments can also have an impact here by promoting locally relevant eGovernment services.

Resources:
Digital Inclusion and Mobile Sector Taxation in Mexico
Mobile Internet Usage Challenges in Asia — Awareness, Literacy and Local Content
Rural Coverage: Strategies for Sustainability
Benefits of Network Competition and Complementary Policies to Promote Mobile Broadband Coverage
Digital Inclusion Report 2014

Digital Identity

Background

The ability to prove that you are who you say you are and have this information authenticated when interacting with the state or private companies is critical to accessing basic services such as healthcare, education and employment, as well as exercising voting rights or benefiting from financial services. However, the World Bank estimates that in the developing world alone around two billion people lack an official identity. These people are predominantly the poorest and most excluded members of society.

This 'identity gap' is both a symptom of slow economic development and a factor that makes development more difficult and less inclusive. The problem is particularly stark when it comes to birth registration, with UNICEF figures showing that one child in three doesn't have a legal identity simply because their birth wasn't registered.

Programme Goals

Both the UN Convention on the Rights of the Child and the Sustainable Development Goals highlight the need to address the lack of birth registration and to provide legal identity to huge numbers of people in developing nations.

The GSMA Digital Identity programme is working with mobile operators and a broad range of stakeholders to establish mobile as a scalable platform for digital identity registration and authentication. World Bank research in sub-Saharan Africa indicates that more than half of the population lacks an official identity, yet more than two-thirds of residents in the region have a mobile phone. These figures highlight the huge potential of mobile to bridge this 'identity gap'.

Public Policy Considerations

When births go unregistered at the national level or people lack official documentation, there is the potential for millions of citizens to be denied access to government, banking and other vitally important services. With children being born into an increasingly connected world and mobile use growing rapidly in developing nations, it makes sense for governments to support innovative and scalable solutions that can provide trustworthy digital identity and authentication to their citizens. Mobile identity is a powerful platform that can help governments achieve this aim and accelerate the move towards digital public services in both developed and developing countries.

However, an enabling regulatory environment needs to be put in place if mobile is to deliver formal identity and authentication to the millions of people who are currently unregistered, and ultimately drive improved social, political and economic inclusion. To help create the right environment, governments need to ensure there is consistency between the different legal and regulatory instruments that affect the management of digital identity, and work to break down any legal, policy and regulatory barriers that may inhibit the roll out of mobile identity services.

Governments also carry a responsibility to foster and help create the trusted environment within which mobile identity operates. The creation of a digital identity plan that acknowledges the central role of mobile in the digital landscape can help here, but governments should also engage with mobile operators, key stakeholders and the wider identity ecosystem to help drive interoperability and innovation.

Resources:
GSMA Digital Identity programme website
GSMA Personal Data website
Mobile Connect — a secure universal log-in solution
Case study: Mobile Birth Registration in sub-Saharan Africa
Report: Mobile Identity — Unlocking the Potential of the Digital Economy

Disaster Response

Background

Mobile networks, and the connectivity they provide, are increasingly identified as a lifeline in disasters due to their ability to support critical communication between humanitarian agencies, affected populations and the international community.

The power of mobile was evident in the aftermath of the 2010 Haitian earthquake, which saw a proliferation of new coordination and response strategies that were built around mobile platforms.

Since then, experiences in the Philippines with Typhoon Haiyan, the Ebola crisis in West Africa and the recent earthquake in Nepal, have continued to provide examples of the critical importance of access to communication and information for populations affected by disaster and crisis.

As the role of mobile in disaster preparedness and response continues to grow, and as the ecosystem becomes more complex, there is a need for a better understanding of how the global mobile communications community can support continued access to communication and information in the wake of disaster.

Programme Goals

The GSMA Disaster Response programme works with mobile operators to determine how they can improve preparedness and network resilience before disaster strikes, and help affected citizens and humanitarian organisations following a crisis.

Through research and engagement with mobile and humanitarian stakeholders, the GSMA is working to define and share best practices and create a robust, coordinated disaster response mechanism for the mobile industry.

This work has culminated in the Humanitarian Connectivity Charter, launched in early 2015. The Charter represents a set of shared principles and activities focused on strengthening access to communication and information for those affected by crisis in order to reduce the loss of life and positively contribute to humanitarian response.

Mobile operators who have signed the Charter currently represent over one billion subscribers in over 35 countries.

Public Policy Considerations

The GSMA has developed a set of recommendations for governments, regulatory bodies and mobile operators to follow during times of crisis.

The key elements of these recommendations are:

- Governments, along with relevant multilateral agencies, and operators should agree a set of regulatory guidelines that can be adopted to best respond to and recover from an emergency.

- The guidelines should set out unambiguous rules and clearly defined lines of communication between all levels of government and operators in emergency situations.

- The guidelines should provide operators with flexibility to adjust to unforeseen circumstances rather than insisting that rules designed for non-emergency situations apply no matter what the circumstance.

- The guidelines should help improve communication and coordination among various government entities involved in responding to an emergency and facilitate a timely and efficient response.

Resources:
GSMA Disaster Response
Humanitarian Connectivity Charter
Dewn: Dialog's Disaster and Emergency Warning Network
Disaster Response: Mobile Money for the Displaced
Business As Usual: How AT&T deals with Natural Disasters
GSMA Guidelines on the Protection of Privacy in the Use of Mobile Phone Data for Responding
to the Ebola Outbreak

Mobile Agriculture

Background

In emerging markets, the agricultural sector not only provides the most employment but is also one of the main contributors to gross domestic product. However, agricultural yields are on average a third lower in developing countries than in developed nations. As a result, farmers in developing countries receive minimal income leaving them stuck in a cycle of poverty, while global food demand is increasing and placing added pressure on the agricultural sector in general. To break this cycle and increase productivity, these farmers need access to information and financial services as well as improved supply chain and market links.

With mobile phone penetration in the developing world growing fast and expected to reach 56 per cent by 2020, mobile operators are uniquely placed to provide relevant services to rural populations. Mobile is the only global infrastructure that can reach huge numbers of agricultural workers and provide the information, financial services and connections to other agricultural businesses that they require in order to boost output and income.

This opportunity is enhanced by the fact that agricultural workers with mobile connectivity will account for nearly half of the total labour force in South Asia and sub-Saharan Africa by 2020. Furthermore, evidence of the beneficial effects of mobile agricultural services is strong. Data from GSMA-supported projects suggests that users of mobile agriculture information services are 30 per cent more likely to grow new crops, use new seeds or new agricultural practices and consequently 39 per cent more likely to experience increased income than those who do not use the services.

Programme Goals

The GSMA mAgri programme works with mobile operators, the development community and agricultural organisations to facilitate the creation of scalable, replicable and commercially sustainable agricultural mobile services. Since its inception in 2009, it has supported six projects across Asia and Africa and reached over 4.7 million farmers.

With support from the UK Government, under its mNutrition Initiative, the GSMA also launched an mAgri Challenge Fund in February 2014. The fund provides risk capital, targeted consultancy and support on service design and content creation to six mobile operator-led projects across South Asia and sub-Saharan Africa to help launch and scale mobile agriculture services. The initiative is on-going and aims to reach 2 million agricultural workers by 2017.

Public Policy Considerations

In some cases, the national Ministry of Agriculture has been important for the success of information-based mAgri services. One example is where organisations linked to the Ministry of Agriculture have provided validation for the content that mobile network operators (MNOs) send to farmers.

However, there are also some challenges that need to be addressed, such as:

Kenya and Tanzania. The Meteorological Departments in both markets have blocked mobile operators from using private weather information, referencing the government's monopoly on this type of data. This kind of barrier hinders the uptake and value proposition of mobile agriculture solutions and needs to be addressed. Independent weather forecast providers have also pointed to the difficulty in gathering the essential historical data used to boost the accuracy of their weather prediction models. Considering the vulnerability of small-scale farmers to climate change, it's critical for service providers to have affordable access to weather data so they can produce highly localised forecasts. These localised forecasts enable farmers to make informed decisions when planting, fertilising and spraying their crops.

India, Bangladesh and Pakistan. There is an increasing number of government-led initiatives that negatively affect the uptake of information-based mAgri services across South Asia. These initiatives are usually introduced as anti-spam measures or to combat non-transparent service charges. In India, the telecoms authority TRAI ruled that customers need to double-confirm their subscription to a service, with the second confirmation having to take place through a third party. In Bangladesh, the regulatory commission BTRC has implemented a similar directive. Operators are now banned from auto-renewing mobile subscriptions to any value added service (VAS). In Pakistan, the industry has seen similar regulatory changes though some operators have benefitted from a more flexible framework.

On the whole, though, these regulations represent a big entry barrier for mAgri services and create further challenges for the mobile for development sector as it tries to reach out to the next 1.8 billion people in developing markets. Affected operators and the GSMA are very concerned about these directives and believe alternative solutions should be explored imminently.

Resources:

GSMA mAgri website
Agri VAS: market opportunity and emerging business models
Women in Agriculture: A Toolkit for Mobile Services Practitioners

Mobile For Development Utilities

Background

The rapid expansion of GSM networks means mobile now has further reach than the electricity grid and piped water networks in most emerging markets. While mobile networks have grown at an astonishing rate of 11 per cent per year since 2000, energy and water access lags behind with yearly growth figures of between one and two per cent. The result is a widening gap between access to mobile and access to utility services. In fact, by 2013 mobile networks covered more than 643 million people without access to electricity and more than 262 million people without access to clean water.

This lack of access to energy and water networks has a profound impact on people's lives. For example, according to figures from charity WaterAid, poor sanitation takes the lives of over 1,400 children per day. And poorer people living off the electricity grid in emerging markets often end up relying on expensive and harmful energy sources, such as kerosene, that suffer from fluctuating prices. As a result, a middle class family in Europe can pay less for energy than a poor family in a country such as Bangladesh.[1]

However, by leveraging the enormous reach of mobile — as well as innovative mobile technologies and services, including Machine to Machine (M2M) communication and mobile money — the industry is well positioned to help bring the life-changing benefits of energy and clean water access to huge numbers of people in emerging markets.

Programme Goals

Challenges to providing universal access to energy, water and sanitation services include last mile distribution, operation and maintenance costs, as well as payment collection.

The GSMA Mobile for Development (M4D) Utilities programme focuses on the opportunity for the mobile industry to leverage its network technology and infrastructure to help solve these challenges in emerging markets.

The programme was established in 2013 thanks to funding from the UK Government Department of International Development. The programme has also launched the M4D Utilities Innovation Grant Fund, which aims to accelerate the development of promising mobile technologies and business models that target improved access to energy and water services.

The key goals of the programme include:

- Supporting the Innovation Fund grantees and their mobile operator partners to help them deliver on the promise of their trials.

- Demonstrating the commercial viability of improving energy and water access through the use of innovative mobile technologies.

- Driving further industry interest and support for increasing access to energy and water services through mobile technology.

Public Policy Considerations

Governments should recognise and support the role mobile can play in improving access to energy and clean water in emerging markets. Mobile technologies are increasingly becoming a key strategic element of the models energy and water service providers use to support service delivery.

For example, many energy and water providers employ mobile M2M technology to support the delivery of their services. Through the use of M2M technologies, water pumps can be monitored remotely and repair call-outs triggered automatically when a fault occurs, reducing down time. Governments should ensure that taxation levels on M2M connections are set at appropriate levels to encourage these types of innovative solutions.

Equally, several companies offering home solar power kits in emerging markets rely on mobile money to make these kits affordable to low-income populations via Pay-As-You-Go financing. Governments should ensure supportive regulation is in place to allow mobile money services to thrive and continue to sustainably provide these much-needed affordable financing schemes.

Furthermore, in developing markets, affordability is critical to increasing the use of mobile phones and associated services such as mobile money. Mobile-specific taxes raise barriers to mobile phone ownership and usage. Governments can play a key role here by ensuring consumers don't face higher taxes on mobile handsets and services than on other goods and services.

[1]   GSMA, Sustainable Energy and Water Access through M2M Connectivity (2011)

Resources:
GSMA Mobile For Development Utilities website
Mobile for Development Utilities Innovation Fund
GSMA Digital Inclusion website

Mobile Health

Background

Providing access to quality healthcare at a sustainable cost is a global issue and a matter of the highest national priority for many governments. The advent of mobile health (mHealth) services offers the public health sector a means to dramatically improve access to healthcare by leveraging the ubiquity of the mobile phone, which has far greater reach than traditional health channels. This is particularly true in developing nations.

However, although there are many mHealth services in the market today, few currently demonstrate scale, replication or significant impact. A study by the GSMA that reviewed almost 700 mHealth services showed that less than one per cent of these significantly impact health outcomes. Four key barriers were identified: fragmentation of service delivery, lack of scale across the full reach of mobile networks, limited replication and a misalignment of the value proposition between mobile and health stakeholders.

The GSMA mHealth programme aims to address these barriers and in doing so foster commercially sustainable mHealth services that truly meet public health needs.

Programme Goals

The GSMA mHealth programme is currently funded by UK Aid and aims to boost maternal and child health via mobile solutions that promote improved nutrition. It has a target of reaching one million mothers by August 2018 across eight markets: Ghana, Malawi, Mozambique, Nigeria, Rwanda, Tanzania, Uganda and Zambia.

As each of these countries has already seen a proliferation of mHealth services, the programme's emphasis is on identifying high-potential mHealth services and helping them to achieve scale, rather than providing grant funding to stimulate the development of new services.

There are three key areas of focus:

- **Research.** The programme's research focuses on identifying priority areas for intervention in each of the target countries, determining knowledge of — and attitudes towards — mHealth among consumers, and identifying the mHealth and mobile partners who are best positioned to achieve scale.

- **Content development.** By engaging with key global and local players in the field of nutrition, the programme encourages the creation of market-specific and culturally-sensitive mHealth content.

- **Industry engagement.** The programme works closely with health and mobile players across both the public and private sector to ensure that services not only become commercially sustainable, but also deliver positive public health outcomes.

Public Policy Considerations

Use cases for mHealth solutions are varied, from mobile services designed for basic phones to sophisticated medical devices with embedded SIMs that collect and transmit patient data back to healthcare providers. As such, there are a wide range of potential regulatory touch points. Clear policy and regulation for mHealth is essential to ensure safety, promote confidence among end users and healthcare professionals, and provide industry with the certainty needed for it to invest in innovation and bring new products and services to market.

Regulatory themes that are of specific interest in emerging markets include:

- **Consent and data protection.** Building trust through suitable approaches to securing consent for data collection and then subsequent protection of that data once it has been collected is important globally, but is often particularly sensitive in developing markets. Frequently, there is a fear of social stigma if information on an individual's diagnosis is inappropriately shared.

- **Systems and interfaces.** In developed countries there has been a proliferation of different standards and systems around mHealth, which often make integration difficult. In contrast, the situation in emerging markets means there is a unique opportunity to define standards that promote interoperability and enable scalability.

Policy themes are more globally applicable, and include:

- **Patient empowerment.** Developing policies that appropriately promote user autonomy and drive mHealth adoption.

- **Reimbursement.** Moving towards reimbursement schedules that reward health outcomes and support innovation.

- **Implementation.** Establishing government programmes that address market barriers, build evidence for the benefits of mHealth and encourage the implementation of mHealth systems and services.

Resources:
mHealth Regulation Impact Assessment: Africa
mHealth Country Feasibility Reports: Nigeria, Ghana, Malawi, Tanzania, Zambia & Mozambique
The Use of Mobile to Drive Improved Nutrition Outcomes: Successes and Best Practices from the mHealth Industry
Catalysing mHealth Services for Scale and Sustainability in Nigeria

Mobile Money

Background

In developing countries, 2.5 billion people are 'unbanked' and have to rely on cash or informal financial services, which are typically unsafe, inconvenient and expensive. However, over one billion of these people have access to a mobile phone. This provides the basis for mobile money, whereby mobile technology is used to deliver convenient and affordable financial services to the underserved.

With mobile money, customers can convert cash to and from electronic value (i.e., e-money), and they can use mobile money to perform transfers or make payments. Banks that rely on traditional 'bricks-and-mortar' infrastructure struggle to serve low-income customers profitably, particularly in rural areas. However, mobile operators have large airtime distribution networks that can be used to provide customers with a network of mobile money agents who perform cash-in and cash-out transactions. Large mobile operators in developing countries typically have 100 to 500 times more airtime reseller outlets than all of the banks' branches put together.

Mobile money has already proven to be viable and sustainable. As of July 2014, there were 245 mobile money services in 88 countries serving more than 61 million active users. At least nine countries now have more mobile money accounts than bank accounts, and 44 countries have more mobile money outlets than bank branches.

Programme Goals

The GSMA Mobile Money programme helps mobile money services achieve scale by identifying and sharing benchmark data, operational best practices and approaches to cross-service interoperability, as well as cultivating positive regulatory environments.

Public Policy Considerations

There are many reasons for governments to encourage digital financial inclusion among their citizens. It contributes to economic growth, it offers convenience and consumer protection, and it reduces the vulnerability of a country's financial system by lowering the risks caused by the informal economy and widespread use of cash.

Mobile money services depend on a regulatory framework that embraces innovation, allowing a new class of financial services providers to sustainably provide digital payment and transfer services. Risks posed by licensed non-bank mobile money providers can be successfully mitigated by requirements that safeguard funds entering the system and ensure customers can cash out electronic value on demand.

An open and level playing field is required, allowing banks as well as non-bank providers to offer mobile money services.

Mobile money reduces the risk of money laundering and terrorist financing, as electronic transactions can be monitored and traced more easily than cash.

Interoperability should not be mandated. In such a young industry, service providers and policymakers should work together to understand different models of mobile money interoperability, including the benefits, costs and risks. The role of the policymaker is to facilitate dialogue between providers, ensuring that interoperability brings value to the customer, makes commercial sense, is set up at the right time, and regulatory risks are minimised.

Resources:
GSMA Mobile Money for the Unbanked
GSMA MMU Deployment Tracker
GSMA 2014 State of the Industry report
GSMA Mobile Money Regulatory Guide
GSMA Mobile Money: Enabling Regulatory Solutions
GSMA The Kenyan Journey to Digital Financial Inclusion
GSMA Enabling Mobile Money Policies in Sri Lanka — The Rise of eZ Cash

Women and Mobile

Background

Mobile phones provide distinct benefits to women, including helping them to feel safer, more independent and more connected. Mobiles also improve women's access to educational and employment opportunities.

However, women are currently under-represented in terms of ownership and usage of mobiles. According to a recent 2015 study[1] commissioned by the GSMA, over 1.7 billion females in low- and middle-income countries do not own a mobile phone. Even those women who do own a mobile tend to use it less frequently and intensively than men, especially for more sophisticated services such as mobile internet and mobile money.

This gender gap can be attributed to a number of factors including the cost of handsets and services, network coverage, concerns around security and harassment, as well as a lack of technical literacy. Social norms are also an issue and can delay – or even prevent – a woman from acquiring a mobile phone and related services.

Programme Goals

The GSMA Connected Women programme works towards greater inclusion of women at all points in the mobile industry chain, to help ensure they can take advantage of the many socio-economic benefits that mobile delivers. It has a specific focus on closing the gender gap in mobile connectivity and the use of mobile money services.

The programme aims to equip mobile network operators and their partners with the knowledge needed to take action to reduce the gender gap in these areas and overcome barriers to women's use of mobile phones. It also focuses on the greater inclusion of women as leaders in the mobile industry.

Public Policy Considerations

Policymakers and regulators can adopt many strategies to ensure women are not excluded from the benefits of mobile. For example, it is important to ensure appropriate policy and regulation is in place to lower cost and access barriers for customers. This can be achieved by reducing mobile-specific taxes, supporting voluntary infrastructure sharing among licensed operators, and releasing sufficient spectrum at affordable cost.

Furthermore, governments can consider strategies for increasing mobile and digital skills through changes to school curricula or the introduction of training programmes. It may also be appropriate to address harassment over mobile phones and mobile internet through awareness campaigns or legal and policy frameworks.

Data on women's mobile phone access and usage, and on ICT more broadly, is also not widely available or tracked in many low- and middle-income countries. Without data, policymakers and the mobile industry cannot make informed decisions to help increase women's access to, and use of, mobile phones. To address this, policymakers can consider options to track mobile access and use by gender, along with other ICTs, in national statistics databases.

Women are also under-represented in the technology sector as employees and leaders. This is important as the technology sector is a high-growth field which is important to countries' innovation, connectedness and competitiveness in global markets. Women today compose 40 per cent of the global workforce and account for more than half of university graduates, yet we see only three to five per cent of senior positions in technology being held by women.[2]

Developing and supporting policies or schemes to address this under-representation is important as it has a measurable economic cost. For example, according to a 2013 European Commission survey on women in ICT, organisations in Europe that have women in senior management positions generate a 35 per cent higher return on equity, while female employment overall provides an annual economic boost of €9 billion.

[1] GSMA Connected Women. Bridging the gender gap: Mobile access and usage in low- and middle-income countries (2015)

[2] MacLeod Consulting, Implications of the ICT Skills Gap for the Mobile Industry (2013)

Resources:
GSMA Connected Women Website
Report: Bridging the gender gap: Mobile access and use in low- and middle-income countries
Report: Accelerating digital literacy: Empowering women to use the mobile internet
Report: Accelerating the digital economy: Gender diversity in the telecommunications sector

Mobile Initiatives

Innovation and investment by the mobile industry continues to have an enormous impact on the lives of billions of people around the world. Mobile doesn't just deliver connectivity, it empowers people through an ever-growing range of mobile-enabled services. Currently there are 3.5 billion unique mobile subscribers worldwide, but this figure is set to grow to 4.6 billion by 2020, representing 60 per cent of the world's population.

The GSMA leads several programmes that are shaping the continued growth and development of the sector. From embedded SIM technology for Internet of Things applications to mobile identity and payment solutions, these initiatives are laying the foundations of an increasingly connected, mobile world.

Each of the initiatives covered on the following pages has its own public policy considerations, and relates to one or more of the public policy topics presented in this handbook.

Future Networks

The strategic importance of Internet Protocol (IP) to future mobile networks is clear and embracing this future is vital for mobile operators as they compete to win and retain customers. Moving to all-IP based infrastructure and services enables operators to deliver a broader, deeper communications portfolio — incorporating voice, data, video and messaging services.

With the increasingly widespread deployment of Long-Term Evolution (LTE) networks, the move to global interconnected IP-communication services such as Voice over LTE (VoLTE), Video over LTE (ViLTE) and Rich Communication Services (RCS) is accelerating at a rapid pace. Through its Network 2020 programme, the GSMA is working with leading operators and equipment vendors to further accelerate the launch of IP-based services around the world.

The mobile industry is also laying the groundwork for the transition to fifth generation (5G) technology. Building on the achievements of 4G, future 5G networks will help the mobile industry capture the huge opportunity presented by the Internet of Things, usher in an era of even faster mobile broadband and pave the way for 5G-optimised services, which may include support for exciting technologies such as tactile Internet, virtual reality and enhanced broadcast services.

The mobile network of the future will also be more energy efficient. Mobile network operators remain overly dependent on fossil fuels to power generators at off-grid mobile base stations. The GSMA is assisting mobile operators with energy assessments and recommendations for using renewable energy sources to reduce operating costs, shrink dependence on diesel fuel and reduce carbon emissions in the provision of mobile service.

5G — The Path to the Next Generation

Background

Mobile telecoms has had a phenomenal and transformational impact on society. Starting from the earliest days of first-generation analogue phones, every subsequent generational leap has brought huge benefits to societies around the world and propelled the ongoing digitisation of more and more segments of the global economy. The mobile industry is now preparing to embark on the transition to fifth generation (5G) technology, which will build on the achievements of 4G while also creating new opportunities for innovation.

A range of industry, research, academic and government groups across the globe are working to define the technology for 5G. The next generation mobile technology will need to provide higher throughput, lower latency and higher spectrum efficiency.

Between now and 2020, the year when 5G is expected to become commercially available, the mobile industry will continue to take steps towards achieving these goals by evolving existing 4G networks. However, despite these enhancements to 4G, there is still a need for 5G to meet the demands of future services and platforms.

Currently, there are three key areas of focus for 5G development and innovation:

Internet of Things (IoT). There is a need for 5G to capture the huge opportunity presented by IoT. Conservative estimates suggest that by 2025 the number of IoT devices will be more than double the number of personal communication devices. As the ecosystem grows, the mobile industry will be expected to support bespoke services across industry verticals and develop next-generation services that are not achievable with 4G networks.

Mobile broadband. With each generational leap in mobile technology there is a natural progression to faster and higher-capacity broadband. Mobile broadband services using 5G technology will need to meet and exceed customers' expectations of faster and more reliable access.

5G-optimised services. Superior speed and reduced latency will see 5G nurture new services that cannot be supported on existing 4G networks. Some of the services being considered include tactile Internet, virtual/augmented reality and remote control of vehicles and robots. Broadcast services are also expected to flourish under 5G.

The GSMA aims to play a significant role in helping to shape the strategic, commercial and regulatory development of the 5G ecosystem. This will include areas such as the definition of interconnect in 5G, as well as the identification and alignment of suitable spectrum bands. Once a stable definition of 5G is reached, the GSMA will work with its members to identify and develop commercially viable 5G applications.

Public Policy Considerations

The GSMA regards 5G as a set of requirements for future mobile networks that could dramatically improve the delivery of mobile services and support a variety of new applications. The mobile industry, academic institutions and national governments are currently actively investigating what technologies could be used in 5G networks and the types of applications these could and should support. The speed and reach of 5G services will be heavily dependent on access to the right amount and type of spectrum.

Additional new spectrum will be required for 5G services in order to deliver enhanced capabilities including new usage scenarios. To ensure 5G services provide good coverage that extends beyond small urban hotspots, it will be important to make sure that there is sufficient spectrum available for this important purpose (i.e., sub-1 GHz spectrum). Progressive refarming of existing mobile bands should be both possible and permitted to accommodate future 5G usage, as well as to maximise spectrum usage efficiency.

The GSMA believes that three key frequency ranges are currently worthy of consideration for different 5G deployment scenarios: Sub-1 GHz, 1-6 GHz and above 6 GHz. Exclusive licensing remains the principal and preferred regime for managing mobile broadband spectrum in order to guarantee quality of service and network investment. However, the licensing regime in higher frequency bands, such as above 6 GHz, could be more varied than in previous mobile technology generations to suit more flexible sharing arrangements.

Resources:
GSMA Understanding 5G: Perspectives on future technological advancements in mobile
GSMA 5G Spectrum Policy Position

IP Communication Services

Background

IP communications is increasingly recognised as a natural evolution of core mobile services, and therefore a basic requirement of doing business in the future. The IP Multimedia Subsystem (IMS) has emerged as the preferred technical means for transferring core mobile operator services to an all-IP LTE environment due to its flexibility, cost-effectiveness and support for IP services over any access medium. With over 40 per cent of the world's mobile network operators having now launched an LTE network, and LTE coverage currently exceeding a quarter of the world population, the industry is now in a realistic position to make a global, interconnected IP communications network a reality. IP communications is comprised of Voice over LTE (VoLTE), Video over LTE (ViLTE), Voice over WiFi (VoWiFi) and Rich Communication Services (RCS).

- **VoLTE** using IMS technology is recognised as the industry-agreed progression of voice services. VoLTE offers an evolutionary path from circuit-switched 2G and 3G voice services and includes a range of enhanced features for customers, such as high-definition audio quality and shorter call connection times. As of November 2015, there were 36 VoLTE services commercially available in 23 countries.

- **ViLTE** will enable operators to deploy a commercially viable, person-to-person video calling service that will revolutionise the way customers communicate with each other. Like VoLTE, it is based on IMS technology.

- **VoWiFi** allows operators to offer secure voice calling over WiFi. As of November 2015, there were 13 VoWiFi services commercially available in eight countries.

- **RCS** marks the transition of messaging and voice capabilities from circuit-switched technology to an all-IP world, leveraging the same IMS capabilities as VoLTE and ViLTE. RCS incorporates messaging, video sharing and file sharing enriching the communication experience of consumers. As of November 2015, RCS was being offered by 45 mobile operators in 33 countries.

The GSMA, via its Network 2020 programme, is working with leading operators and equipment vendors to accelerate the launch of IP-based services around the world. The work of the Network 2020 programme covers the development of specifications, assisting operators with the technical and commercial preparations for service launches and resolving technical and logistical barriers to interconnect.

Public Policy Considerations

To support the exponential growth in IP traffic, large-scale investments in network capacity are required. Financing such investments depends on predictability and the existence of a stable regulatory environment. Where such an environment exists, future communications capabilities that are operator-led can be well aligned with the regulatory requirements related to mobile telecommunications, and mobile network operators have the systems in place to ensure compliance.

Open standard. VoLTE, ViLTE, VoWiFi and RCS are currently specified, through a process of industry collaboration, as open industry standards for IP-based calling, messaging, file and video sharing services, generically based on IMS technology.

Interconnect. VoLTE, ViLTE, VoWiFi and RCS support interconnection of these services between customers on two different mobile networks.

Lawful intercept. Mobile network operators are subject to a range of laws and licence conditions that require them to be capable of intercepting customer communications and to disclose this information to law enforcement agencies on their request. The specifications for IP communications are being developed so they support the capabilities needed to meet lawful interception obligations.

Resources:
Report: Building the case for an IP-communications future
All-IP Business Guide
Report: The Value of Reach in an IP World

Mobile Energy Efficiency

Background

Mobile network operators (MNOs) spend approximately $15 billion on their annual energy use. Therefore, it is no surprise that energy efficiency is a strategic priority for them globally. As mobile use continues to grow, so does the demand for energy, particularly by the network equipment used by the mobile industry. At the same time, mobile technology plays an important role in enabling energy efficiency in other sectors, and as a result has significant impact right across the global economy, including in greenhouse gas reduction.

Mobile's Green Manifesto 2012, published by the GSMA, outlines the positive effects that mobile operator initiatives are having in energy and carbon management, as well as the progress being made around mobile's role as an enabler for energy efficiency. The report also found that there were 26 million mobile machine-to-machine (M2M) connections worldwide in 2012, with these helping to reduce greenhouse gas emissions by an estimated three million tonnes of carbon dioxide emissions (CO_{2e}) annually.

However, the report also showed that mobile has the potential to support much greater emissions savings — at least 900 million tonnes of CO_{2e} in 2020, which is 1.7 per cent of the global 2020 greenhouse gas emissions forecast by the International Energy Agency in its 'business-as-usual' scenario. In part, these savings will come from a huge increase in the number of mobile M2M connections, which are forecast to reach around 3.5 billion by 2020. Carbon-reducing applications for M2M technology are wide-ranging – from improving fleet operations and making utilities more efficient to enabling smart cities and monitoring retail stock levels remotely.

Programme Goals

To help mobile operators reduce their energy costs and greenhouse gas emissions, the GSMA's Mobile Energy Efficiency (MEE) programme offers two services to MNOs: MEE Benchmarking and MEE Optimisation.

MEE Benchmarking is a management tool that helps MNOs measure and monitor the relative efficiency of their radio access networks, identifying under-performing networks and quantifying the potential efficiency gains available, typically around ten to 25 per cent across a mobile network operator's portfolio.

MEE Optimisation is a follow-on service that uses the MEE Benchmarking results combined with site audits and equipment trials, first to analyse the costs and benefits of specific actions to reduce energy and emissions, and second to roll out the most attractive solutions. The service is run in partnership with a third-party vendor or systems integrator, and the GSMA has assembled a group of technology partners to undertake these projects.

Public Policy Considerations

Through the MEE service, the GSMA is contributing to the Global eSustainability Initiative Energy Efficiency Inter-Operator Collaboration Group (GeSI EE-IOCG) which is working to develop common ICT industry standards for energy efficiency. In addition, the GSMA is collaborating with the European Commission, the International Telecommunication Union (ITU) and the European Telecommunications Standards Institute (ETSI) on standardisation, including methodologies to assess environmental impact. For example, the GSMA contributed to ETSI standard ES 203 228 on assessing the energy efficiency of mobile networks, which was published in April 2015.

Government can play an important role in the promotion of green initiatives by supporting the development of these robust methodologies for assessing environmental impact and backing the creation of ICT industry standards for energy efficiency.

In addition, energy efficiency has considerable wider economic and social benefits. For example, it reduces the power drawn from the grid, therefore making more electricity available for the rest of society.

As it can take several years for operators to recoup their investments in green solutions, and in particular renewable energy solutions, it is important that policymakers and regulators ensure there is clarity and stability around policies and incentives in these areas. Several countries have already put incentivised regulations in place. For example, Bangladesh, Indonesia, Pakistan and Uganda have all introduced tax and fiscal incentives for renewable energy players or green telecoms.

Results from several of the GSMA's MEE Optimisation projects clearly show the value of investing in energy efficiency and renewables. For example, mobile operator Warid Telecom Pakistan worked with telecoms energy solution provider Cascadiant to trial a range of equipment on Warid's cell sites. This project demonstrated significant energy savings of between 30 and 60 per cent and when rolled out across Warid's network the annual savings are expected to be more than $6 million and 19,700 tonnes of CO_{2e}.

Similarly, another MEE Optimisation project saw mobile operator Telefónica Germany team up with the then Nokia Siemens Networks' Energy Solutions to look at how the operator could cut its energy costs. The project identified potential savings in energy costs of €1.8 million per annum with a payback timeframe of considerably less than three years and emissions savings of 4,000 tonnes of CO_{2e}. Many of the recommendations have now been implemented.

Resources:
GSMA Mobile Energy Efficiency
GSMA Mobile's Green Manifesto 2012
Report: Mobile Energy Efficiency — An Overview

Voice over Long Term Evolution

Background

Consumers expect seamless carrier-grade voice services from mobile operators, irrespective of the type of technology used.

Since the introduction of digital mobile technologies in the early 1990s, carrier-grade public mobile voice services have been delivered via the circuit-switched capabilities of 2G and 3G networks.

To keep pace with growing demand, mobile operators are now upgrading their networks using a fourth generation IP-based technology standard called Long Term Evolution (LTE). LTE networks incorporate a new carrier-grade voice capability called Voice over LTE (VoLTE) that offers an evolutionary path from circuit-switched 2G and 3G voice services. VoLTE includes a range of enhanced features for customers such as high-definition audio quality and shorter call connection times.

Some operators now have LTE networks that offer full national coverage and are using VoLTE for voice calls. Other operators still only have partial LTE network coverage.

In most markets it will take a number of years to phase out 2G and 3G networks and fully migrate customers to LTE-based networks and services. For voice services, the transition is facilitated by the fact that VoLTE has been designed to support the seamless hand-over of calls to and from 2G and 3G networks.

VoLTE has a number of characteristics that distinguish it from Internet-based voice services. These include carrier-grade call quality and reliability, and universal interconnection with other 'carrier-operated' voice services across the globe. By contrast, the majority of Internet-based voice services are not managed for service quality and may be restricted to closed user groups.

In some jurisdictions, interconnection of carrier-grade mobile voice services is unregulated and carried out pursuant to a range of different commercial agreements. In other jurisdictions, regulated mobile call termination rates apply. These rates typically use a time-based charging mechanism and their levels are set using a number of different cost-orientated methodologies.

Public Policy Considerations

Voice over Long Term Evolution (VoLTE) is a carrier-grade mobile voice service, making it distinct from other internet-based voice services.

Carrier-grade mobile voice services have a number of specific characteristics. For example, the use of mobile phone numbers from national numbering schemes means that customers can make calls to, or receive calls from, any other phone number in the world. Carrier-grade mobile voice services also assure end-to-end service quality and reliability by using dedicated network capacity (technically known as bearers).

VoLTE is an evolution of carrier-grade mobile voice services that have historically been provided using the circuit-switched capabilities of 2G and 3G networks. As such, regulators should not apply additional, or specific, regulations to VoLTE services.

In markets where mobile voice call termination is subject to regulatory control, the same approach should be adopted for VoLTE, with a single rate applied across 2G, 3G and 4G/LTE voice call termination.

Resources:
Article: ECN Magazine, 'VoLTE: What makes voice over IP 'carrier-grade'?'

Internet of Things

The Internet of Things (IoT) is set to have a huge impact on our daily lives, helping us to reduce traffic congestion, improve care for the elderly, create smarter homes and offices, increase manufacturing efficiency, and more.

IoT involves connecting devices to the internet across multiple networks to allow them to communicate with us, applications and each other. It will add intelligence to devices that we make use of on a daily basis and in turn deliver positive impacts to both the economy and broader society.

We are set to see rapid growth in IoT over the coming years. According to GSMA Intelligence, the number of cellular Machine-to-Machine connections is expected to have reached just under one billion by 2020. However, this will still represent a small portion of the overall market as Juniper Research predicts that by 2020, the total number of IoT devices will have grown to 38.5 billion.

The GSMA, through its Connected Living programme, is encouraging the development of the nascent IoT ecosystem by working to define industry standards, promote interoperability and encourage governments to create a supportive environment that will speed the growth of IoT globally.

Encouraging the Growth of IoT

Background

The Internet of Things (IoT) promises to deliver a huge range of benefits to citizens, consumers, businesses and governments. Referring to machines, devices and appliances of all kinds that are connected to the internet through multiple networks, the IoT has tremendous potential to shrink healthcare costs, reduce carbon emissions, increase access to education, improve transportation safety and much more.

Through its Connected Living programme, the GSMA aims to accelerate the delivery of these types of connected devices and services, and thereby enable a world in which consumers and businesses enjoy rich new services, connected by an intelligent and secure mobile network.

The IoT market is already developing at a rapid pace. According to figures from GSMA Intelligence, by the end of 2014 the number of cellular IoT connections had reached nearly 250 million, with that figure set to soar to just under one billion by 2020. Understandably, governments and regulators are increasingly interested in how they can capture the benefits of the IoT and channel them to their citizens.

However, IoT business models, markets and services are fundamentally different from traditional telecoms services, such as voice and messaging. In most cases, IoT services have a closed user group and the customers are not typically end users of the service, but businesses that need to be able to roll out IoT solutions globally. Also, IoT services are characterised by a significantly lower average revenue per connection than traditional voice and messaging services.

Therefore, if governments are to create a supportive environment for the IoT, they must recognise these differences when considering policy and regulatory frameworks. This means policy and regulation should be flexible, balanced and technology-neutral to ensure they support large-scale deployments and encourage investment.

Public Policy Considerations

There is huge potential for the IoT to transform economies and societies, but the technologies and ecosystem that support IoT are still at an early stage of development. If governments are to realise the significant socio-economic benefits that IoT can deliver, they must foster an investment-friendly and technology-neutral environment that will allow it to grow and flourish.

Governments can achieve this by putting in place policies that provide the right incentives for growth and innovation. They can also lead by example through the adoption of IoT solutions in the public sector or by funding research and development programmes.

As the IoT ecosystem is composed of a large number of diverse players, policy frameworks must be based on the fair regulation of equivalent services. Regulatory clarity is also hugely important to give service providers and IoT device manufacturers the confidence to make the necessary investments in this emerging technology for it to achieve global scale.

Governments and regulators can play a significant role here too, by supporting and promoting interoperable specifications and standards across the IoT industry. This is important to the future growth of the IoT, as interoperable platforms and services reduce deployment costs and complexity, facilitate scalability and enable consumers to enjoy intuitive connected experiences.

As the IoT is projected to grow hugely in the coming years, governments also need to adopt a flexible framework for both licensed and unlicensed spectrum to ensure mobile operators can deploy the most appropriate technology mix.

The IoT presents significant opportunities for data-driven innovation to achieve economic, social and public policy objectives and improve people's daily lives. However, for this to happen, data protection and privacy legal frameworks need to be practical, proportionate and applied consistently to all parties in the IoT value chain. This will help create a climate of trust between industry and end users.

Resources:
GSMA Connected Living website
GSMA Connected Living Tracker

Global Deployment Models for IoT

Background

The Internet of Things (IoT) is ushering in an era where unprecedented numbers of devices will become connected all around the globe. The scale and reach of this machine-to-machine (M2M) connectivity will allow new services to develop that can help societies make more efficient use of resources across a range of industries and sectors including healthcare, agriculture, transportation and manufacturing.

However, if governments and societies are to tap into these benefits, companies operating within the IoT ecosystem will need to be able to deploy their services on a global, rather than local, scale. It is only by following global deployment models that the nascent IoT industry can pass on to consumers the benefits they get from economies of scale for service delivery.

Global approaches to service deployment have a number of advantages. For example, they accelerate the speed and quality of deployment and also drive down the cost of servicing smaller, local markets where the creation of a bespoke local service would be uneconomical. Furthermore, they help guarantee the delivery of a consistent, high-quality experience to the end user.

Mobile operators are already taking the lead in supporting global service launches in early market categories such as automotive, health and consumer electronics. With the emergence of new products in adjacent categories, including healthcare and wearable devices, the importance of being able to support large-scale, global deployments is likely to increase.

There are a range of different global deployment models operators can choose to use, including M2M international roaming, the embedded SIM technology developed by the GSMA's Connected Living programme, or a hybrid of the two. New deployment models may also emerge in the future.

The choice of deployment model may depend on a number of factors, such as:

- The particular needs of the mobile operator, IoT service provider and end user.

- The scale and geographical footprint of the deployment.

- The type of IoT application and its unique service requirements.

- The device lifetime and its accessibility.

Public Policy Considerations

The IoT has the potential to bring substantial social and economic benefits to citizens and businesses through more efficient use of resources, the creation of new jobs and services, increases in productivity and improvements in service delivery.

However, IoT business and distribution models are very different to those used to deliver traditional telecoms services, such as voice and messaging. Typically, they are global in nature, with elements of the value chain spread across various countries and regions.

The great diversity in the range of services on offer and the partners involved in IoT, as well as this geographical spread in the value chain make it hugely important for the industry to be able to develop and select the most suitable deployment models for different types of IoT services. This is why policymakers and regulators should avoid regulation that tries to steer the industry towards a one-size-fits-all approach to deployment. Instead, governments should encourage innovation in IoT deployment models and understand that operators will be required to adopt flexible commercial and technical solutions in different countries and regions around the world.

Governments can support the global nature of the IoT market in other ways, such as by backing interoperable platforms and services to reduce deployment costs and complexity, ensuring that all players in the IoT market are operating on a level regulatory playing field, and working together across jurisdictions to ensure consistency and clarity on legal, data protection and privacy regulation.

Resources:
PriceWaterhouseCoopers Realising the benefits of mobile-enabled IoT solutions
GSMA Understanding the Internet of Things

mAutomotive — Connected Cars

Background

The integration of mobile communications into vehicles is changing people's relationship with the car. Increasingly, drivers and passengers are able to obtain real-time information about their trip (such as weather conditions or traffic flow data) and enjoy car-appropriate infotainment (such as internet radio and video services for passengers). Large-scale deployments of connected car solutions already exist in many parts of the world, and the variety of services is growing significantly.

Mobile network operators, which have traditionally provided connectivity for vehicle services, are beginning to move up the value chain, offering extended connectivity support (e.g., applications management), expanded core assets (e.g., customer service management, billing systems and fraud management) and sector-specific services such as telematics service provision, disaster recovery and datacentre hosting.

Through its Connected Living programme, the GSMA is actively engaging with vehicle manufacturers, mobile network operators, SIM vendors, module makers, and the wider automotive ecosystem to facilitate the development of current and future connected automotive solutions.

The primary platform for these activities is the Automobile Special Interest Group (Auto SIG). This group was established by the GSMA with the aim of promoting dialogue across all stakeholders in the automotive ecosystem and looks to find innovative ways that mobile technology can be leveraged by the automotive sector.

Currently a key area of focus is the GSMA's Embedded SIM Specification. This provides a single mechanism for the remote provisioning and management of machine-to-machine (M2M) connections, allowing 'over-the-air' provisioning of an initial operator subscription, as well as subsequent changes of subscription from one operator to another.

The Embedded SIM specification has global backing (from operators, SIM suppliers and a wide variety of equipment and vehicle manufacturers) and offers a number of key advantages that make it particularly suitable for mAutomotive applications:

- It is live and commercially available now from leading global mobile operators.

- It offers the same level of security achieved today by traditional SIMs.

- It reduces risks of tampering, as the SIM is soldered into the vehicle.

- It reduces the need for a mechanical SIM holder and slot.

Public Policy Considerations

There are substantial benefits that mAutomotive applications bring to consumers, including making driving safer and providing infotainment to passengers. Furthermore, mAutomotive applications can deliver enormous socio-economic benefits, but it is important policymakers understand that most of these applications are at a nascent stage of development.

Governments can help encourage the development of the mAutomotive ecosystem by maintaining the right incentives for growth and innovation, promoting research and development programmes and supporting efforts around service and network interoperability.

Many mAutomotive applications have distinct characteristics. Some of these are common to other M2M applications, such as a longer 'device' lifetime and the need for services to operate globally. Others are sector specific, including regulations covering the security and emergency elements of mAutomotive solutions.

It is important that policymakers and regulators appreciate and understand these differences and implement policies that allow for global business models to develop within the mAutomotive sector, while ensuring that policies apply consistently to all players in the value chain. Policies should also be technology and service neutral, and support a level playing field for all players in the industry. This will help build trust and confidence in mAutomotive solutions among end users.

Currently, security and emergency regulations have been introduced in three locations: Europe, Russia and Brazil.

- In Europe, the regulations relate to eCall, an in-vehicle emergency call system that automatically triggers an emergency call in the event of a severe road accident. The proposed legislation requires all new vehicles sold in the EU to be eCall-ready by March 2018.

- The GSMA is involved in two EU-led task forces for eCall: Lifecycle Management of the SIM and the Periodic Inspections Tests. The former relates to the provisioning of the in-car SIM (from its activation through to defining the events that trigger the SIM 'end-of-life') and the latter concerns the testing processes that will be put in place to ensure that all cars sold in the EU by March 2018 have a fully functioning eCall system.

- In Russia, ERA GLONASS has similar goals to eCall and extends to insurance reconstruction and dangerous goods transport services, while Brazil's SIMRAV project focuses on reducing vehicle theft and lowering vehicle insurance rates through mandatory fitment for stolen vehicle location services.

Resources:
GSMA mAutomotive
Report: Connected Car Forecast Next Five Years
White paper: Split Charging and Revenue Management Capabilities for Connected Car Services
White paper: Connecting Cars — Tethering Challenges

Mobile Health and IoT

Background

The pressures on healthcare systems have never been greater, due to factors including rising expectations, ageing populations and, particularly in emerging economies, the combined challenges of infectious disease and increasing incidence of chronic illness. Mobile health solutions provide an opportunity to help healthcare providers deliver better, more consistent and more efficient healthcare, increasing access to health services and empowering individuals to manage their own health more effectively.

According to a 2015 report by PWC, mHealth could save over one million lives in sub-Saharan Africa over the next five years and the use of Internet of Things (IoT) technology in healthcare could reduce healthcare costs by €99 billion in the European Union and add €93 billion to the region's gross domestic product by 2017.

Many mobile health propositions have gained acceptance and are being more widely adopted. The market is developing, and this growth is accompanied by a rapid increase in the number of solutions that potentially offer new modalities of care. Greater consideration is therefore being given to the policy and regulatory frameworks that will govern their promotion and use.

Public Policy Considerations

Use cases for mHealth solutions are varied, from medical devices that collect patient data to applications that deliver health services and information. As such, there are a wide range of potential regulatory touch points.

Although significant progress has been made over the last few years, there is an ongoing need for clarity in policy and regulation related to mHealth to ensure safety, promote confidence among patients and healthcare professionals, and provide industry with sufficient certainty to bring new products and services to the market.

Policy themes include:

Patient-centre healthcare. Developing policies that promote patient-centred care and user autonomy to help drive mHealth adoption.

Access. Promoting initiatives to integrate mHealth services into healthcare systems and care pathways to encourage the development of reimbursement schemes that reward health outcomes and support innovation.

Implementation. Building evidence and establishing government programmes to enable large-scale implementations of mHealth solutions.

Systems, interfaces and interoperability. Promoting interoperability and standards that support scalability and a plug-and-play experience.

Regulatory themes include:

Medical devices. Developing and implementing clear and proportionate regulatory frameworks that aim to ensure patient safety while stimulating innovation.

Data protection. Ensuring an appropriate regulatory framework is in place for data protection and privacy is of key importance. Regulatory measures should be proportionate and facilitate the use of data in creating patient-centered and sustainable healthcare systems.

Consumer use – Telehealth

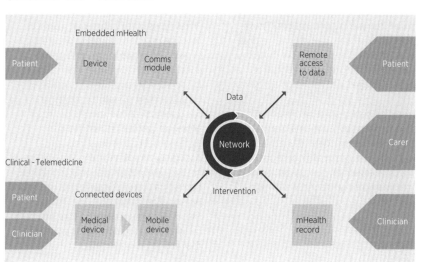

Source: PA Consulting Group

Resources:
GSMA Response to European Commission Green Paper on mHealth
Europe: Joint Statement of the Healthcare Coalition on Data Protection
GSMA Position Paper: Medical Device Regulation
PWC Realising the benefits of mobile-enabled IoT solutions
GSMA Report: Potential of Mobile Health Solutions to Address Chronic Disease Challenges
GSMA Video: Transforming Healthcare with Embedded Mobile
PA Consulting Group: Policy and Regulation for Innovation in Mobile Health
PWC Report: Socio-Economic Impact of mHealth, European Union
PWC Report: Socio-Economic Impact of mHealth, Brazil and Mexico

Privacy and Data Protection for IoT

Background

The Internet of Things (IoT) offers significant opportunities and potential for data-driven innovation to achieve economic, social and public policy objectives, and ultimately improve people's daily lives. For example, the IoT will enable a raft of new applications and services that will empower consumers to monitor their health, manage their energy consumption and generally benefit from smart home and city solutions. These applications have the potential to drive a range of positive outcomes including improved traffic management, lower pollution levels and healthier lifestyles.

Many IoT services will be designed to create, collect or share data. Some of this data (for example, data about the physical state of machines or weather conditions) may not impact on consumers' privacy and as a result won't be considered personal data.

However, IoT services aimed at consumers are likely to involve the generation, distribution and use of detailed data about those consumers. For example, a smart home appliance may use data about a person's eating or exercise habits to draw inferences about that person's health and steer them towards healthier lifestyles or develop a profile based on their shopping habits to offer them personalised money-off vouchers.

These types of IoT services and devices have the potential to impact people's privacy and may be subject to general data protection and privacy laws. Where IoT services are provided by mobile operators they will also be subject to telecommunications-specific privacy and security rules. Nevertheless, as consumer IoT services gain in popularity, more consumer data will be created, analysed in real time and shared between multiple parties across national borders. Therefore companies throughout the IoT ecosystem have a responsibility to build trust among consumers by ensuring their privacy is respected.

Public Policy Considerations

To realise the opportunities that the IoT offers, it is important for consumers to trust the companies who are delivering IoT services and collecting data about them. The mobile industry's view is that consumer confidence and trust can only be fully achieved when users feel their privacy is appropriately respected and protected.

There are already well-established data protection and privacy laws around the world. Where these data protection regulations and principles exist, they can also be applied to address privacy needs in the context of IoT services and technologies. It is vital that governments apply these frameworks in ways that promote self-regulation and encourage the adoption of risk management-based approaches to privacy and data protection.

Most importantly, protections should be practical, proportionate, and designed into IoT services (privacy by design) to encourage business practices that provide transparency, choice and control for individuals.

IoT services are typically global in nature and a mobile operator is often only one of many parties in a delivery chain that may include a host of others, such as device manufacturers, search engines, online platforms and even the public sector. Therefore, it is key that that privacy and data protection regulations apply consistently across all IoT providers in a service and technology-neutral manner. This will help ensure there is a level playing field for all industry players so they can focus on building trust and confidence for end users.

Resources:
U.S. Senate Subcommittee: Respect for privacy vital for growth of the IoT
GSMA The Impact of the Internet of Things
GSMA Privacy Design Guidelines for Mobile Application Development

Personal Data

Digital content, services and interactions have become a part of daily life for billions of people, driven by expanding access to broadband and increasingly affordable connected devices. Personal data and user authentication are requisite elements of being online — users must identify themselves to be able to access their accounts and subscriptions, to make purchases, and so on.

The digital economy is based on trust. Interactions — whether they be social, commercial, financial or intellectual — require a proportionate level of trust in the other party or parties involved. Without such trust, users will find other ways to browse, bank and buy.

Currently, user authentication is inconsistent and inconvenient for users, and people are forced to keep track of numerous login names and passwords. Meanwhile, identity theft is on the rise. Failure to address these problems will create barriers to market digitalisation and social inclusion.

To this end, the mobile industry is developing a consistent and standardised set of services for managing digital identity, putting mobile at the heart of digital identity management. With mobile operators' unique advantages such as the SIM card, strong registration processes, network authentication and fraud detection, mobile operators have the ability to provide sufficient authentication to enable consumers, businesses and governments to interact in a private and secure environment.

The GSMA is working with mobile network operators and mobile ecosystem players, as well as governments, banks and retailers, to help roll out mobile identity solutions. The association is also working with industry standardisation bodies such as the Open ID Foundation to ensure support and interoperability for global standards.

Together, mobile operators will bring digital identity solutions to the market with scale, offering a seamless consumer experience, consistent technology and low barriers to entry across the digital identity ecosystem.

Advantages of mobile operators in providing a digital identity service

The mobile device
Ubiquitous, personal and portable; sensitive to location and capable of being disabled and locked.

The SIM card
Real-time strong authentication; encryption for storing certificates and other secure information.

Know your customer (KYC) standards
Strong registration and fraud-detection processes in place.

Robust regulatory requirements
Established systems to handle personal data safely.

Customer service
Sophisticated customer care processes and billing relationships.

Verified subscriber data
Ready for mobile identity.

Flexibility to innovate
Ability to add consumer functionality such as 'add to bill' or 'click to call'.

Mobile Connect

Background

At Mobile World Congress 2014, the GSMA with the support of leading mobile operators unveiled the Mobile Connect initiative, a digital identity solution that offers a safe, seamless and convenient consumer experience, a consistent user interface and low barriers to entry across the digital identity ecosystem — thereby enabling global scale.

GSMA Mobile Connect is a secure universal solution that simplifies consumers' lives. By matching the user to their mobile phone, Mobile Connect allows them to log in to websites and applications quickly without the need to remember passwords and usernames. It is safe, secure and no personal information is shared without permission. This opens up a range of opportunities for both mobile operators and consumer-focused service providers to build a rich suite of offerings for their customers, while ensuring the user's private and confidential information is kept safe.

- For consumers, Mobile Connect enhances user privacy, reduces the risk of identity theft and simplifies the login experience for a range of services. It does this by leveraging the established data handling processes of the operators and inherent security of the SIM for authentication and identification. With Mobile Connect, the user is authenticated through their mobile phone, rather than through personal information, making logging in safer and more secure. With a streamlined, secure login, consumers have easier access to e-government, retail and banking services, among others, without the need to remember additional passwords.

- For service providers, Mobile Connect offers the advantages of an improved consumer experience (including reduced drop-off rates when signing on to new services), lower cost of managing credentials, and validation of important consumer attributes such as age.

The standards-based GSMA Mobile Connect solution utilises the OpenID Connect protocol, offering broad interoperability across mobile operators and service providers, further ensuring a seamless experience for consumers. Mobile Connect can also provide different levels of security, ranging from low-level website access to highly-secure, bank-grade authentication. Mobile Connect promises to make passwords a thing of the past.

Programme Goals

Initially, the focus is on achieving a consistent approach across the mobile industry to provide authentication services such as seamless login. This means that the consumer chooses to use Mobile Connect as their digital identity solution when they sign up for a new service with a provider (or add it later), and the provider then queries the mobile operator for the credentials of the consumer. As a result, the consumer can remain anonymous to the service provider, while the service provider benefits from a better way to manage credentials and the ability to provide the consumer with a more convenient user experience for its services.

Public Policy Considerations

Mobile identity services inevitably involve multiple devices, platforms and organisations that are subject to differing technical, privacy and security standards. Some governments are already using mobile technology as a key enabler to deliver digital identity services in their digital plans. However, to achieve wide adoption and the greatest impact on the economy, a number of public policy issues must be addressed:

- Identify and assess existing legal, regulatory and policy challenges and barriers that affect the development of mobile identity services.

- Leverage best practice to foster wide-scale mobile identity services and transactions.

- Engage with mobile operators and the wider ecosystem to facilitate interoperability and innovation.

Governments should create a digital identity plan that acknowledges the central role of mobile in the digital landscape. The mobile industry is committed to working with governments and other stakeholders to establish trust, security and convenience in the digital economy.

The mobile industry has a proven track record of delivering secure networks and has developed enhanced security mechanisms to meet the needs of other industry and market sectors. The implementation and evolution of these security mechanisms is a continuous process. The mobile industry is not complacent when it comes to security issues and the GSMA works closely with the standards development community to further enhance the security features used to protect mobile networks and their customers.

In summary, mobile operators, with their differentiated identity and authentication assets, have the ability to provide sufficient authentication to enable consumers, businesses and governments to interact in a private, trusted and secure environment and provide more secure and convenient access to services.

Resources:
GSMA Personal Data Programme
GSMA Mobile Connect
Mobile Identity: A Regulatory Overview
Case study: Norwegian Mobile Bank ID: Reaching Scale Through Collaboration
Case study: Swisscom Mobile ID: Enabling an Ecosystem for Secure Mobile Authentication

Digital Commerce

Mobile operators around the world are working with retailers, loyalty scheme providers and equipment vendors to roll out mobile services for digital commerce. For example, some mobile operators are offering a mobile wallet — a specialist application that can store digital versions of payment cards, loyalty cards, vouchers, tickets and other items normally found in a physical wallet.

These services allow individuals with handsets that are equipped with Near Field Communications (NFC) technology to touch their handset against a point of sale terminal to redeem vouchers, make a payment, collect loyalty points and more.

This type of interaction will fundamentally change how people conduct financial transactions. The GSMA is working with regulators to develop and support the ecosystems needed to roll out these types of sophisticated digital commerce propositions around the world.

Near Field Communications

Background

Businesses and consumers are looking to digital commerce to provide flexible and efficient transaction services across a range of sectors, including retail, transport, financial services, online and advertising.

Near Field Communications (NFC) is a wireless technology that can transfer information between two devices within a few centimetres of each other. There are already more than 300 SIM-based NFC launches, including 60 operator-led commercial services around the world.

Mobile operators, as they seek to harness the potential of NFC, are engaging with the relevant actors in their markets, including local and national governments, transportation bodies, banks, retailers and other stakeholders. In some cases, mobile operators are forming joint ventures with other operators and banks. In others, they are engaged in partnerships based on business models that incentivise all the actors in the NFC value chain.

Operators have opportunities to partner with card schemes and use their tokenisation platforms to dramatically simplify the link up between banks and SIM-based mobile payment services. Tokenisation is the process of reducing security risks by substituting sensitive data with a non-sensitive equivalent that has no exploitable meaning or value. The substituted data, or token, is used as a reference to map back to the sensitive data through the tokenisation platform.

If operators are able to use a card scheme's tokenisation platform, this will allow banks to have access to all operators in a market through a single integration process with that card scheme. The same integration also gives the banks access to other competing channels that have recently emerged, such as ApplePay. Without these types of partnerships, banks would have to carry out an integration process with each individual operator, which is time consuming and costly.

The GSMA is focused on driving tokenisation deployments and establishing clear guidelines for their implementation. The SIM is one of the channels being used as a digitised payment instrument as part of these efforts. However, the GSMA is also looking to extend the tokenisation proposition beyond the SIM to customers without an NFC SIM. The GSMA is also actively investigating the opportunities arising from the adoption of the European Commission's Payments Services Directive 2. These elements will be critical to the widespread adoption of NFC, enabling people around the world to benefit from NFC services, regardless of their operator network or device type.

Public Policy Considerations

SIM-based NFC handsets can provide robust security features, such as PIN numbers to access services and strong authentication techniques (such as a digital signature or one-time password) to protect the mobile wallet. Moreover, the mobile operator can activate and deactivate services over the air if the phone is lost or stolen, and reinstall services once a new phone is provisioned. The SIM also complies with international security standards and is tamper resistant.

SIM-based NFC reduces the need for cash and plastic cards, leading to operational efficiencies and cost savings. In some cases, SIM-based NFC could also reduce fraud, increase the number of customers who can be served at one time, help track inventory and facilitate value-added services, such as automatic coupon redemption.

Mobile NFC services have the potential to lower barriers to entry for smaller service providers. This could lead to increased competition, more choice for consumers and reduced prices.

Resources:
GSMA Report: The New Mobile Payment Landscape
GSMA HCE and Tokenisation for Payment Services discussion paper
GSMA Report: Socio-Economic Benefits of SIM-Based NFC
GSMA Report: Mobile and Online Commerce, Opportunities provided by the SIM
GSMA White Paper: Mobile NFC in Retail
GSMA Report: The Value of Mobile NFC in Transport 2014

Business Environment

Governments have a responsibility to create a business environment that supports innovation and allows industry to thrive so it can have a positive social and economic impact. The mobile sector is highly dynamic, so flexible, light-touch regulation is essential. The market is best able to drive and shape the industry's evolution, as highly prescriptive regulatory policy cannot keep pace with the swift advance of mobile technologies, services and consumer demand.

One example is found in the current asymmetry that exists between the regulatory requirements placed upon mobile operators versus those of the internet players that provide IP-based voice and messaging services.

The mobile sector is among the most intensely regulated industry sectors, subject not only to common rules governing consumer protection and privacy, but a raft of sector-specific rules related to interoperability, security, emergency calls, lawful intercept of customer data, universal service contributions and more. It is also one of the most heavily taxed sectors around the world, facing a variety of industry-specific taxes, levies and fees.

Base Station Siting and Safety

Background

Mobile services are a key enabler of socio-economic development, and achieving ubiquitous access to mobile services for citizens is a major government policy objective in most countries. Mobile operators often have roll-out obligations in their market area to ensure widespread national coverage.

To deliver continuous mobile coverage in dense urban areas and across rural expanses, mobile network operators must build and manage an array of base stations — free-standing masts, rooftop masts and small cells — equipped with antennas that transmit and receive radio signals, providing voice and data services to their customers in the area.

A variety of requirements and conditions, including electromagnetic field (EMF) exposure limits, must be met to secure permits for base-station deployment. Requirements can be defined at the local, regional and national levels, even though the local authority (e.g., the municipality) is typically the point of referral. The process in some countries leads to significant delays and cost variances.

Debate

What planning permission processes should governments implement to avoid undue delay in infrastructure installation?

What reference point should be used by governments to define safe EMF exposure limits?

How can a balance be struck between national objectives for mobile connectivity by citizens and the decisions of municipalities?

Can processes be streamlined for approval of small cell antennas?

Industry Position

Governments that enable mobile network investment and remove barriers to the deployment of network infrastructure will accelerate the provision of mobile services to their citizens.

By defining explicit, nationally consistent planning approval processes for mobile base stations, governments can avoid lengthy delays in network deployment. We support mechanisms that reduce bureaucratic inefficiencies, including exemptions for small installations, colocations or certain site upgrades, 'one-stop shop' licensing procedures and tacit approval. Governments can lead by example by improving access to government-owned land and buildings.

Base-station exposure guidelines should be aligned with international standards as recommended by the World Health Organization (WHO) and International Telecommunication Union (ITU). Additional restrictions related to environmental impact should be avoided.

Infrastructure costs place a high threshold on entry into the mobile sector. If policies are short-sighted, and if taxes and licence fees are not in keeping with actual market dynamics, then operators may not have the means, or the will, to roll out new technologies and to reach rural areas. Such policies delay the social and longer-term economic benefits experienced by citizens.

Resources:
Report: Base Station Planning Permission in Europe
World Health Organization: Electromagnetic Fields
Federal Communications Commission (USA): Wireless Infrastructure Order
GSMA: Arbitrary Radio Frequency Exposure Limits – Impact on 4G Network Deployment
GSMA Infographic: Mobile Networks for a Better-Connected World
GSMA: LTE Technology and Health

Facts and Figures

Radio Frequency Policies for Selected Countries

Country	RF Limit at 900MHz (W/m²)	Requirement for RF licensing	Exemptions or simplified procedures for...	Location restrictions	Consultation during siting process
Australia	4.5	Compliance declaration	Small antennas, changes	None	Yes
Brazil	4.5	Approval	–	50m [a]	Local
Canada	2.7[b]	Approval	Small antennas, changes	None	Yes
Chile	4.5/1	Approval	Small antennas, changes	>50 [c]	Yes
Egypt	4	Approval	–	20m [d]	No
France	4.5	Approval	Small antennas, changes	Voluntary, to minimise exposure [e]	Local
Germany	4.5	Approval	Small antennas, changes	None	Yes
India [f]	0.45	Compliance declaration	–	None nationally, local variation	No
Italy	1/0.1	Approval	Small antennas	Lower limits [g]	Yes
Japan	6	Approval	Small antennas	None	Local

Country	RF Limit at 900MHz (W/m²)	Requirement for RF licensing	Exemptions or simplified procedures for...	Location restrictions	Consultation during siting process
Kenya	4.5	Compliance declaration	Changes	None	Yes
Malaysia	4.5	Approval	Small antennas	None	Yes
Netherlands	4.5	Compliance declaration	Small antennas, changes	None	Yes
New Zealand	4.5	Compliance declaration	Small antennas, changes	None	Local
Kingdom of Saudi Arabia	4	Compliance declaration	–	None	No
South Africa	4.5	Compliance declaration	–	None	Local
Spain	4.5	Approval	Small antennas, changes	None	Local
Turkey ʰ	1.5	Approval	–	None	Local
United Kingdom	4.5	Compliance declaration	Small antennas, changes	None	Yes
United States	6	Approval	Small antennas, changes	None	Local

a   50m around hospitals, schools and homes for old people

b   Proposal under public consultation

c   ICNIRP with lower limit in urban areas and in 'sensitive areas'

d   Not within 20m of schools and playgrounds

e   Recommendation to minimise exposure in schools, day-cares or healthcare facilities located within 100m

f   Adopted ICNIRP in 2008 and changed to 10% of ICNIRP on 1 September 2012

g   Lower limit in playgrounds, residential dwellings, schools and areas where people are >4 hours per day

h   One installation; total exposure must not exceed ICNIRP 1998

Competition

Background

Mobile phones are the most widely adopted consumer technology in history. A large part of this success can be attributed to how competition in the mobile industry has helped drive innovation.

The rise of the digital economy and explosive growth in smartphone adoption have brought innovation and disruption to traditional mobile communications services. These changes are also impacting existing policy frameworks and challenging competition policy (which includes government policy, competition law and economic regulation).

Despite the influence that new market dynamics are having on the mobile sector, the industry is still subject to the contradictions of a legacy regulatory system. This has resulted in services that are in competition with each other — such as voice services offered by mobile operators and those offered by internet players — being regulated differently.

These differences can be seen in how economic regulation (ex-ante) and competition law (ex-post) are applied to the sector. For example, a regulator's jurisdiction may be limited to the telecommunications sector, and not extend to internet players. As a result, regulators often fail to take wider market dynamics into account during the evaluation and decision-making process. Equally, a failure to understand the complex value chain can affect how competition law is applied.

The end result is that mobile operators are currently caught between the two worlds and consumers may not receive the full benefit of these competitive markets.

Debate

How should markets be defined in the Digital Age?

How can standard competition tools be applied in the Digital Age?

Are traditional significant market power (SMP) access remedies still appropriate?

Industry Position

The mobile industry supports competition as the best way to deliver economic growth, investment and innovation for the benefit of consumers. Excessive regulation can stifle innovation, raise costs, limit investment and harm consumer welfare due to the inefficient allocation of resources, particularly spectrum.

Regulators and competition authorities should recognise fully the additional dynamic competition in the Digital Age. Internet players adopt new and different business models to offer services to customers. Examples include advertising-supported services that make use of sophisticated internet analytics. Regulators and competition authorities need to understand these models, and map their competitive impact before imposing regulatory obligations or competition law commitments. Otherwise, services that are in competition with each other may end up being regulated differently. For example, players that adopt traditional, better understood business models may find themselves subject to enhanced scrutiny.

Accounting for these new types of competitors when conducting market assessment reviews may show that there is a much greater level of competition in communication services markets than is currently recognised by regulatory and competition authorities. This type of analysis could demonstrate the potential for regulatory policy goals to be achieved through competition law, with the result that ex-ante regulation could be lessened, or may no longer be needed. As a result, a degree of deregulation of licensed providers may be justified.

All competitors providing the same services should be subject to the same regulatory obligations, or absence of such obligations. The principle of 'Same Service, Same Rules' can be achieved through the use of deregulation described above and of horizontal legislation that should progressively replace industry, technology or service-specific rules.

Resources:
GSMA Competition Policy in the Digital Age – A Practical Handbook For Policy Makers (2015)
Report: Challenges for Competition Policy in a Digitalised Economy

Comparing the digital value chain to the current system of competition law and regulation

In the digital value chain many internet players are competing head to head with mobile operators in the provision of communication and entertainment services to end users. For example, some freemium services compete directly with services where upfront prices are charged and can impose a formidable competitive constraint on mobile operators seeking to charge a price to the end user.

This can be seen clearly in the impact that instant messaging (IM) continues to have on SMS and circuit switched voice services (see chart below). The new competition pressures are not limited to messaging services. For example, users can also choose to switch to Wi-Fi for data or voice services instead of using a mobile network.

The justification for regulating mobile networks needs to be considered in light of these facts. It is also important to consider the possible knock-on effect on the ability of mobile operators to invest in their networks if the services they offer are regulated more tightly than the services with which they compete, distorting competition and artificially reducing market opportunities for mobile operators.

Mobile operators, if they so wished, could adopt a range of different business models, similar to those used by internet players but, sometimes due to regulation (and data protection regulation in particular), this is not always possible.

More generally, the mobile industry is currently subject to both telecom-specific regulation and the traditional application of competition rules, while competing internet players are not subject to ex-ante regulation and could even escape competition law scrutiny altogether due to the characteristics of their business model. As an example, the way in which turnover (i.e., revenue) thresholds are often determinative of merger control scrutiny, leads to a situation where mergers involving internet players may not be subject to scrutiny even though the purchasing price (reflecting the market valuation of the target) may be high.

A sector specific regulator can only use the tools it has at its disposal, within the limits of its jurisdiction. If past assessments by a telecoms regulator continue to inform the analysis of this new type of marketplace, there is a risk that larger issues in the digital economy may wrongly be perceived as a telecoms sector-specific issue leading to extra layers of regulation on mobile operators. This is sometimes referred to as Maslow's hammer problem: "it is tempting, if the only tool you have is a hammer, to treat everything as if it were a nail." (Maslow, The Psychology of Science, 1966).

Regulators and competition authorities should fully recognise the additional dynamic competition in the Digital Age when carrying out market reviews — including SMP reviews, competition law investigations and merger control reviews (for a detailed consideration of the latter, refer to the Efficient Mobile Market Structures chapter of this book). Whether in SMP regulation or in competition law, the assessment of market power should take into consideration the dynamics of a market in constant evolution.

In competition law, it is easier to apply traditional categories to telecommunications providers that adopt a traditional business model, including charging an upfront price for services. The argument that where there is no price, there is no market, may lead to the conclusion that there are no competition issues with freemium services.

With respect to selecting the appropriate factors for market reviews in the context of today's dynamic competition, it may also be appropriate to ask whether certain factors should be considered specifically. These include the monetisation of personal data, the existence of multi-sided markets and strong network effects, as well as the coexistence of closed systems alongside open ones.

The traditional approach of competition authorities and telecoms regulators also fails to recognise that consumers of a freemium service may pay a price in other ways, for example by sharing their personal information with a provider of freemium services. We are seeing a competitive process emerging around data usage and this constitutes a new dimension that needs to be considered in market analysis.

Impact of instant messaging on circuit switched voice and SMS volumes

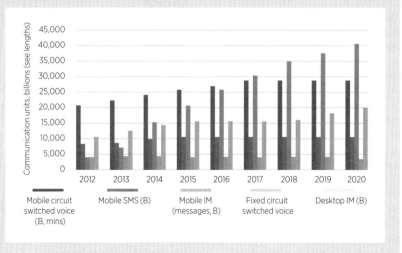

Efficient Mobile Market Structures

Background

From the outset, mobile markets have been characterised by a vibrant, competitive market structure that drives investment and innovation. Traditionally, the main policy tool used to support this market structure has been spectrum licensing. From 2000 onwards, policymakers have licensed an increasing number of mobile network operators in an effort to drive competition and improve market performance.

A number of metrics have been used to measure market performance under this structure, including prices, subscriber growth and investment in network coverage, as well as increased network speeds and quality.

The policy tool of spectrum licensing has led to the current situation where the number of countries with a single mobile provider has reduced from about half of the countries in the world in 2000 to a small number of states representing less than three per cent of the world's population today[1].

The demand from both users and policymakers for high-speed, high-quality, robust and secure mobile networks is great and has driven mobile operators to make large investments in network infrastructure and services. These investments are ongoing as the mobile industry typically follows a ten year (or shorter) technology cycle, so while operators are currently investing heavily in 4G networks, in a few years' time the focus of investment will shift to 5G technology.

The high level of competition in the markets for mobile services has seen the tariffs charged to mobile users fall dramatically, so users now get more for their money. While competition is key when it comes to driving innovation and wider societal benefits, policymakers must avoid creating or maintaining artificial and uneconomic conditions that force prices down to untenable levels that deter operators from investing in their networks.

As a consequence of increased competition in mobile markets, the need for ex-ante regulation in these markets has faded away. National regulatory authorities must therefore recognise the competitive nature of today's mobile markets, avoid interventions aimed at engineering market structures and allow market mechanisms to determine the optimal mobile market structure. At the same time, competition authorities tasked with assessing the impact of proposed mobile mergers must take full account of the dynamic efficiencies (and accompanying wider societal benefits) arising from mobile mergers.

[1] Frontier Economics 'Assessing the case for Single Wholesale Networks in mobile communications', September 2014, p.16.'

Debate

Can mergers between mobile operators bring significant consumer benefits in mobile markets and wider society?

Industry Position

When assessing mobile mergers, policymakers should consider the full range of static and dynamic benefits that can arise from mergers, including price effects, innovation, the use of spectrum and investments over both the short and longer term.

Investment and Quality of Service

- Competition authorities should consider placing greater emphasis on how mergers may change an operator's ability to invest. Growing demand for data services requiring ever increasing bandwidth means constant investment in new capacity and technology is needed.

Positive spill-over effects in the wider economy

- Improvements in digital infrastructures support economic growth by positively affecting productivity across the whole economy.

Greater benefits than network sharing

- Competition authorities have often argued that network sharing represents a preferred alternative to mergers. While the pro-competitive nature of network sharing agreements can only be assessed on a case-by-case basis, it is worth noting that network sharing agreements are not always feasible between the merging parties due to an asymmetry of assets (such as spectrum holding) or a different deployment strategy.

Unit prices

- There is no robust evidence to suggest that four-player markets have produced lower prices than three-player markets in Europe and elsewhere over the past decade.

- Mergers can accelerate the transition between technology cycles in the mobile industry (technology cycles are responsible for significant reductions in unit prices), leading to improvements in quality and driving service innovation.

- As the market moves from voice to data, the global volume growth rate on mobile networks is accelerating. This calls for more concentrated market structures than in the past in order to meet the investment challenge and drive mobile data unit price down so as to keep the demand for mobile data services growing.

Effects of remedies on investments and use of spectrum

- In some cases, if operators are compelled to provide third parties with access to their networks, this could reduce rather than sharpen incentives to invest as a result of the merger, thus significantly reducing benefits to consumers. In addition, in the three cases (Ireland, Germany and Austria) where a network entry option was made available by the European Commission's Directorate-General for Competition, nobody took the option, even though this was arguably offered on favourable terms.

- Remedies that involve reallocating network assets or reserving spectrum for other operators could in some cases deter investment and lead to under-utilised or misused resources.

Resources:
Assessing the case for in-country mobile consolidation
Assessing the case for in-country mobile consolidation in emerging markets

Deeper Dive

Issues with pricing indexes

Competition authorities currently use the Gross Upwards Pricing Pressure Index (GUPPI) to estimate the incentive of the merged company to raise prices. However, GUPPI analysis has proved unreliable when estimating the impact on the unit prices expected to result from mobile mergers. For example, evidence from the recent merger in Austria confirms unit prices did in fact fall and not increase as authorities had anticipated.

GUPPI calculations are based on the flawed hypothesis that only operating costs are taken into account by mobile operators to build retail prices, and that network investment costs are ignored. This is clearly untrue.

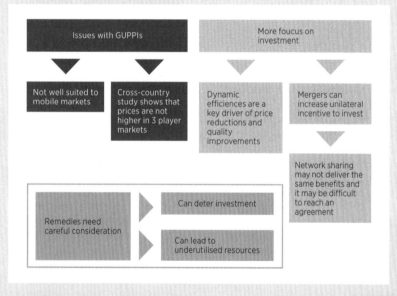

Source: Frontier Economics

Deeper Dive

The relationship between market structure, competition and investments

While the existence of a relationship between market structure and innovation has been well known to economists since the 1950s, the exact nature of this relationship has been, and still is, widely debated.

Competition authorities need to consider carefully potential changes to market structure in order to understand whether a proposed merger will hinder or enhance competition, and identify what corresponding impact it will have on consumer welfare.

In 2005, economists suggested that the relationship between competition and innovation takes the form of an 'inverted U' under which competition has a positive effect on innovation until it reaches a point of inflexion, after which more competition has a negative effect. See 'Competition and innovation: an inverted U relationship', Aghion, Bloom, Blundell, Griffith and Howitt, 2005.

Recent economic theory suggests that while most markets benefit from an increase in the number of competitors, there will come an inflexion point at which adding further competitors will actually reduce market performance.

More recently, economists have been turning their attention to the relationship between competition, investments and innovation in mobile markets, in particular with reference to the impact of consolidation. A recent study analysed changes in mobile market structure covering more than 34 OECD countries over the period between 2002 and 2014. It found that a reduction in the number of players in a market can cause investments to go up significantly following a merger[1]. Another study, by HSBC, found that investments do go up following in-market consolidation[2].

It can also be mentioned that investment in network quality to improve consumer experiences generally leads to decreasing returns. For this reason, the point for which the commercial benefit of a quality increase matches its cost will be higher if the number of customers benefitting from the quality improvement is larger, such as when the market is more concentrated.

[1] Evaluating Market Consolidation in Mobile Communications', Genakos, Valletti and Verboven, Centre on Regulation in Europe (CERRE), 2015.

[2] 'Supersonic', HSBC Telecoms, Media & Technology Global, April 2015

Environment and Climate Change

Background

As mobile use expands, so does the demand for energy, particularly by the network infrastructure. The mobile industry is responsible for a small fraction, less than 0.5 per cent of global greenhouse gas (GHG) emissions, but energy is a significant cost for mobile operators, especially in emerging markets.

An analysis of 65 mobile networks showed that total network energy consumption increased only four per cent from 2010 to 2011, despite considerable growth in mobile traffic and connections. Total energy per unit traffic declined by approximately 30 per cent, and energy per connection declined by three per cent.

The mobile industry's goal is for global GHG emissions per connection to drop by 40 per cent between 2009 and 2020.

The European Union, in particular, is pushing for the information and communication technology (ICT) sector to use detailed carbon accounting to help the EU meet GHG reduction targets.

Debate

In addition to the mobile industry's continued focus on reducing its own emissions, should it also work towards ICT-enabled emission reduction in other sectors? If so, how can governments help?

What is the role of government in using mobile technology to reduce emissions generated by its own public services, for example by promoting green ICT solutions?

Does mandated carbon accounting generate sufficient benefit, when there is no common, agreed methodology?

Industry Position

The mobile industry acknowledges its role in managing greenhouse gas emissions, but also believes governments should encourage mobile machine-to-machine (M2M) communications in sectors where the potential to reduce emissions is greater.

Research has identified the potential for the mobile industry to reduce GHG emissions in other sectors — including transportation, buildings and electrical utilities — by at least four to five times its own carbon footprint. The savings principally come from smart grid and smart meter applications, as well as smart transportation and logistics.

The mobile industry is taking active steps to increase the energy efficiency of its networks and reduce emissions. With mobile network operators spending around $15 billion on energy use annually, energy efficiency and emission reduction are strategic priorities for them globally.

The GSMA's Mobile Energy Efficiency Benchmarking service enables network operators to evaluate the relative energy efficiency of their networks. More than 40 mobile operators have participated in the service, accounting for more than 200 networks and over half of global mobile subscribers.

The GSMA's Mobile Energy Efficiency Optimisation service uses the benchmarking results in conjunction with site audits and equipment trials to analyse the costs and benefits of energy and emission-reduction actions, and roll out the most attractive solutions. A project with Warid Telecom Pakistan and Cascadiant demonstrated potential energy savings of more than $6 million and a reduction of 19,700 tonnes of carbon dioxide emissions per year.

The GSMA's Mobile Energy Efficiency methodology has been adopted in the International Telecommunication Union (ITU) recommendation for environmental impact assessment of ICT networks and services. The GSMA has also contributed to the European Telecommunications Standards Institute's work on developing international standard ES 203 228, which defines an energy efficiency measurement method for base stations.

The Green Power for Mobile programme, a joint initiative of the GSMA and the International Finance Corporation (IFC) — a member of the World Bank Group — promotes the use of renewable and green energies to extend mobile coverage beyond the available grid.

Resources:
GSMA Mobile Energy Efficiency
Mobile's Green Manifesto 2009 and 2012 update
GSMA Green Power for Mobile
GeSI Smarter2030 analysis
Broadband Commission: Task Force on Sustainable Development and the Post 2015 Development Agenda
Broadband Commission: Linking ICT with Climate Action
ITU-T and Climate Change

A Green Power Feasibility Study for Airtel Madagascar

Globally, a 16 per cent increase in off-grid and poor-grid telecommunications sites is expected in the next six years. Adoption of alternative and renewable power generation is necessary for mobile operators to keep operation costs in check and responsibly manage the volume of carbon emissions their networks generate. To this end, the GSMA Green Power for Mobile programme works with mobile operators to provide market analysis and consulting, technical assistance and business model design.

In 2013, the GSMA conducted a green power feasibility study for Airtel Madagascar to demonstrate the technical feasibility and financial viability of green power alternatives to the operator's existing power approach, in order to reduce Airtel's dependence on diesel generators and hence reduce CO_2 emissions. The feasibility study acknowledged a number of challenges faced by the operator, including:

- Poor access to network base stations.

- Low penetration of grid power and high cost of grid extensions.

- High cost of diesel for off-grid base station generators.

- Lack of domestic suppliers for renewable energy and technologies.

- Lack of policy support for renewable energy deployment.

Given these conditions, the GSMA advised Airtel, to implement a hybrid grid-battery approach for its on-grid sites, to reduce dependence on a diesel generator to power the base station. For off-grid sites, three options were identified: extending grid power to the base station, installing a renewable power solution, or implementing a diesel generator and battery combination.

Following the GSMA's site-by-site analysis, Airtel was advised to implement a solar-hybrid energy solution for 147 sites, extend grid power to 48 sites and implement a diesel-battery hybrid for 21 sites. Other recommendations included implementing smart-energy monitoring and equipment-control mechanisms for all sites, and installing smart power-source controls to select the appropriate power source (i.e., solar, grid power, batteries and diesel generator).

Airtel Madagascar has begun implementing the recommended changes, and the GSMA calculates that the operator will reduce its energy bill by over 90% across the 147 sites where a green solution is deployed. In the case of off-grid or poor grid sites, energy costs can constitute as much as 75% of a site's annual operation cost. Airtel Madagascar used to spend approximately $25,000 per year on energy generation and management for one site plus approximately $9,000 covering rent, overhead and battery replacement costs. After the solar-hybrid implementation Airtel's energy generation and operation costs will drop to around $3,000 per site per year.

In addition to the financial advantages of this green energy approach, the environmental outcomes will be considerable when the upgrades are complete:

- A reduction in diesel consumption of 1.12 million litres per year.

- A 75 per cent reduction in diesel generator dependency.

- Green energy solutions offering an average return on investment within 2.25 years.

- Reduced CO_2 emissions by 3,120 tons per year.

- 978,876 kWh per year generated from renewable energy sources.

Gateway Liberalisation

Background

International gateways (IGWs) are the facilities through which international telecommunications traffic enters or leaves a country.

In emerging markets, fixed-line telecoms incumbents were granted monopolies over IGWs, the assumption being that an IGW monopoly allows a country to manage its international charges and, in so doing, enables the incumbent to fund a national network roll out.

Through changes in technology and the deployment of new services such as VoIP, it has become possible to bypass monopoly gateways. Such examples of bypass have significantly increased competition and lowered international prices.

Unfortunately, some countries have levied a new telecommunication specific tax in the form of a surcharge on international inbound traffic (SIIT), which amounts to double taxation for inbound calls.

The presence of monopoly international gateways tends to also inflate the price for mobile roaming services.

In the late 1990s and 2000s, most countries liberalised IGW. By the end of 2013, less than 15 per cent of markets remained monopolies and typically these are very small island nations, or under-developed, troubled states.[1]

Debate

Which structure for international gateways, monopoly or liberalised, best serves a country and its citizens?

The evidence shows that liberalisation actually stimulates investment and that the fear of loss of international revenues is illusory... Combined with the wider economic benefits to a country and its government, IGW liberalisation is a rational and best practice regulatory response to the IGW monopoly.

— GSMA Research report on the Benefits of Gateway Liberalisation, 2007

Industry Position

Competition in international gateway services should be encouraged, as it leads to reduced consumer costs, more international bandwidth and improved quality of service to operators.

IGW liberalisation delivers macroeconomic benefits by lowering the cost of business, ensuring diversity of supply and international competitiveness, attracting investment and increasing connectedness in the global economy.

Countries that have attempted to maintain IGW monopolies are vainly attempting to hold back the tide, as illegal bypass can account for up to 60 per cent of traffic. Although bypass delivers cheap prices to consumers, it does so at the cost of service quality and the risk of service interruption when local services relying on illegal technologies are shut down.

For developing countries to fully participate in a globalised world, their IGWs must be fully liberalised to allow competition and private investment.

By allowing IGW monopolies to operate, governments are faced with significant regulatory and law-enforcement costs to prevent illegal bypass, while losing out on the tax revenue that could be generated by legal services.

Where the liberalisation of an IGW is intended, international best practice suggests that competitive safeguards can be put in place to ensure that the environment evolves in a fair manner. There may be a need to regulate incumbent operators to ensure reasonable access to 'bottlenecks' (such as cable stations, duct work and backhaul), which are under the control of the incumbent.

[1] Arthur D Little research for GSMA 2015

Resources:
GSMA Report: Gateway Liberalisation: Stimulating Economic Growth
GSMA Report: Mobile Taxation: Surcharges on International Incoming Traffic

Infrastructure Sharing

Background

Common in many countries, infrastructure sharing arrangements allow mobile operators to jointly use masts, buildings and even antennas, avoiding unnecessary duplication of infrastructure. Infrastructure sharing has the potential to strengthen competition and reduce the carbon footprint of mobile networks, while reducing costs for operators.

Infrastructure sharing can provide additional capacity in congested areas where space for sites and towers is limited. Likewise, the practice can facilitate expanded coverage in previously underserved geographic areas.

As with spectrum trading arrangements, mobile infrastructure sharing has traditionally involved voluntary cooperation between licensed operators, based on their commercial needs.

Debate

Should regulators oversee, approve or manage infrastructure-sharing arrangements?

What role should governments play in the development and management of core infrastructure?

Industry Position

Governments should have a regulatory framework that allows voluntary sharing of infrastructure among mobile operators.

While it may at times be advantageous for mobile operators to share infrastructure, network deployment remains an important element of competitive advantage in mobile markets. Any sharing should therefore be the result of commercial negotiation, not mandated or subject to additional regulatory constraints or fees.

The regulatory framework of a country should facilitate all types of infrastructure sharing arrangements, which can involve the sharing of various components of mobile networks, including both so-called passive and active sharing.

In some cases, site sharing increases competition by giving operators access to key sites necessary to compete on quality of service and coverage.

Infrastructure sharing agreements should be governed under commercial law and, as such, subject to assessment under general competition law.

Access to government-owned trunk assets should be available on non-discriminatory commercial terms, at a reasonable market rate.

Resources:
GSMA Report: Mobile Infrastructure Sharing
ZDnet: Could Tower-Sharing Be the Solution to Rural Networks' Problems?
ITU: Mobile Infrastructure Sharing
Article: Indus Towers — The India Way of Business

Types of Infrastructure Sharing

Infrastructure sharing can be passive or active. Passive sharing includes site sharing, where operators use the same physical components but have different site masts, antennas, cabinets and backhaul. A common example is shared rooftop installations. Practical challenges include availability of space and property rights. A second type of passive sharing is mast sharing, where the antennas of different operators are placed on the same mast or antenna frame, but the radio transmission equipment remains separate.

In active sharing, operators may share the radio access network (RAN) or the core network. The RAN-sharing case may create operational and architectural challenges. For additional core sharing, operators also share the core functionality, demanding more effort and alignment by the operators, particularly concerning compatibility between the operators' technology platforms.

Infrastructure sharing optimises the utilisation of assets, reduces costs and avoids duplication of infrastructure (in line with town and country planning objectives).

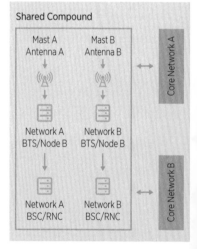

It may also:

- Reduce site acquisition time.

- Accelerate the roll-out of coverage into underserved geographical areas.

- Strengthen competition.

- Reduce the number of antenna sites.

- Reduce the energy and carbon footprint of mobile networks.

- Reduce the environmental impact of mobile infrastructure on the landscape.

- Reduce costs for operators.

Full RAN Sharing   Shared Core Network Elements and Platforms

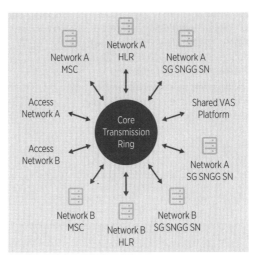

Source: GSMA

Intellectual Property Rights — Copyright

Background

Copyright is the basis for creative industries, collecting societies and artists to earn income from audio and visual work. The original intention of copyright was to encourage the development of new creative work. This is still the case today, but the emergence of the internet as a place for buying, sharing, downloading and streaming content has created challenges for policymakers and stakeholders, including combatting piracy, reforming content licensing and clearly establishing consumer rights.

The debate on buying, sharing, downloading and streaming content — particularly across borders — is bound to continue. However, for some time now it has been clear that the current European Union (EU) copyright provisions from 2001 need to be adapted to the realities of today's digital world. Several initiatives were taken in the past few years, but with the adoption of the European Commission's Digital Single Market (DSM) Strategy in May 2015 modernising copyright law has now become one of the Commission's key priorities for the next five years.

A legislative initiative guaranteeing better online access to digital goods and services across the EU was due to be presented to the Commission by the end of 2015. The focus for this was expected to be on territoriality and portability issues (including unjustified geo-blocking). It was also expected to include a review of the intermediary liability regime to determine whether intermediaries (including mobile operators) should be required to exercise greater responsibility and due diligence.

Views in the debate vary widely. Rights holders advocate strong laws and cooperation of internet service providers and telecom companies in fighting piracy. Civil society organisations defend consumers' fundamental rights (e.g., freedom of expression and access to the internet) and strongly oppose any measures to combat piracy. Collecting societies oppose content licensing reform and defend national licences.

Debate

Should mobile network operators be expected to monitor and address the unlawful use of copyrighted content on their networks?

Is a device levy a legitimate way to compensate artists and publishers for their creative works?

What is the best way for Europe or other regions to enable intellectual property to be used by mobile subscribers in multiple countries?

Industry Position

The mobile industry recognises the importance of proper compensation for rights holders and prevention of unauthorised distribution. However, communications service providers, including mobile network operators and ISPs, should not be held liable for illegal, pirated content on their networks and services, provided they are not aware of its presence and follow certain rules to remove or disable access to the illegal content as soon as they are notified by the appropriate legal authority.

The development of new content licensing models should fall to the rights holders. Obligations on ISPs to monitor piracy should be light touch, if they are employed at all.

Expanding the legitimate content market is crucial when it comes to fighting illegal file sharing. Therefore a key priority should be to reduce existing barriers to cross-border dissemination of content by creating simple, legally sound licensing terms and guaranteeing portability of bought and/or subscribed content, without weakening the territoriality of copyright.

Resources:
Reda Report
The Digital Single Market Strategy
Public Consultation on the review of the EU copyright rules
Directive on collective management of copyright
GSMA consultation response on EU Property Rights Enforcement Directive
Orphan Works Directive

The Economic Importance of Copyright

Copyright industries are defined by the World Intellectual Property Organisation (WIPO) as those industries in which copyright plays an identifiable role in creating tradable private economic (property) rights, and income from the use of these economic rights. This classification defines copyright industries in four groups:

Core industries, which exist to create copyright materials.

Dependent industries, which manufacture equipment that facilitate copyright activity.

Partial industries, which don't create copyright but are dependent on it.

Support industries, which distribute copyrighted material.

The original intention of copyright was to encourage the development of new creative work. It was a system put in place to stimulate incentives for artistic production. Copyright is still a critical foundation for the creative industries, and it is these industries that are most impacted by copyright infringement, in particular commercial-scale piracy, with counterfeiting having a greater impact on the partial copyright industries. Frontier Economics estimates the total value of all counterfeiting and piracy globally to have reached $1,220 billion to $1,770 billion by 2015, with digitally pirated goods alone estimated to account for $80 billion to $240 billion of the total value.

Classification	Example industries
Core copyright industries	Literature, music, theatre, film, video, radio, photography.
Copyright-dependent industries	TV sets, CD players, games equipment, photocopiers.
Partial copyright industries	Household goods, footwear, apparel, museums, libraries.
Non-dedicated support industries	Retailing, transportation, telecommunications.

In the digital economy, copyright continues to perform the critical function of encouraging new works but also has a wider impact, playing a significant role in fostering innovation. The impact of copyright is therefore now much wider than the creative industry alone. Digital technologies, the companies that exploit them and the business models they facilitate are all potentially impacted by copyright.

International Mobile Roaming

Background

International mobile roaming (IMR) allows people to continue to use their mobile device to make and receive voice calls, send text messages and email, and use the internet while abroad.

Telecoms regulators and policymakers have raised concerns about the level of IMR prices and the lack of price transparency, which can cause consumer bill shock.

In the European Union, roaming regulation has been in place since 2007. The latest regulation requires European mobile operators to provide wholesale roaming access services to alternative roaming providers, enabling them to offer competing retail roaming services within Europe. In regulating roaming access in this way, the EU seeks to increase competition, with the aim of removing the need for price cap regulation.

In December 2012, during the revision by the ITU of the International Telecommunications Regulations (ITRs), several governments requested that the revised treaty include provisions on transparency and price regulation for mobile roaming. However, on balance, ITU Member States concluded that roaming prices should be determined through competition rather than regulation, and text was included in the treaty to reflect this approach.

Bill shock and certain high roaming prices have also attracted the attention of international institutions such as the OECD and the WTO. Additionally, regional and bilateral regulatory measures are either in place or being considered in many jurisdictions.

Debate

Some policymakers believe IMR prices are too high. Is regulatory intervention the right way to address this?

What measures can be taken to address concerns about price transparency, bill shock and price levels?

What other factors affecting roaming prices do policymakers need to consider?

Industry Position

International mobile roaming is a valuable service delivered in a competitive marketplace. Price regulation is not appropriate, as the market is delivering many new solutions.

The mobile industry advocates a three-phased strategy to address concerns about mobile roaming prices:

- **Transparency.** In June 2012, the GSMA launched the Mobile Data Roaming Transparency Scheme, a voluntary commitment by mobile operators to give consumers greater visibility of roaming charges and usage of mobile data services when abroad.

- **Removal of structural barriers.** Governments and regulators should eliminate structural barriers that increase costs and cause price differences between countries. These include double taxation, international gateway monopolies and fraud, all of which should be removed before any form of IMR price regulation is considered.

- **Price regulation.** Governments and regulators should only consider price regulation as a last resort, after transparency measures and innovative IMR pricing have failed to address consumer complaints, and after structural barriers have been removed. The costs and benefits of regulation must be carefully assessed, taking into account unique economic factors such as national variances in income, GDP, inflation, exchange rates, mobile penetration rates and the percentage of the population that travels internationally, as well as incidence of international travel to neighbouring countries, all of which have an impact on IMR prices.

The mobile industry is a highly competitive and maturing industry, and one of the most dynamic sectors globally. In the past decade, competition between mobile operators has yielded rapid innovation, lower prices and a wide choice of packages and services for consumers. Imposing roaming regulation on mobile operators not only reduces revenue and increases costs, but it deters investment.

Resources:
GSMA Information Paper: Overview of International Mobile Roaming
Press Release: GSMA Launches Data Roaming Transparency Initiative
GSMA Roaming

Case Study

Roaming Regulation in the EU

Following six years of information requests and public consultation, the first EU roaming regulation was proposed by the European Commission in 2006. The debate centred on the need for retail price controls and the legitimacy of the use of the EU legal framework for the Single Market. The regulation, which came into force on 30 June 2007, obliged operators to introduce a Eurotariff for roaming within Europe as the default roaming plan. The regulation set Eurotariff and wholesale price ceilings following a downward glide path.

This intervention was followed by a second roaming regulation in 2009, which extended and lowered the existing caps on voice calls and extended them to also cover SMS (wholesale and retail) and data transfers (wholesale caps only). It also implemented a number of measures to increase the consistency and transparency of billing for these services, including per-second billing, the cut-off facility on roaming data charges at monthly €50 by default. Customers traveling to another Member State also receive an automated message of the charges that apply for roaming services.

Evidence from innovative roaming offers suggest that market dynamics will deliver roaming prices close to domestic rates in the near future, driven in particular by the move from voice to more price elastic data usage. Nevertheless, from April 2016 regulated retail roaming surcharges cannot exceed the regulated wholesale caps currently in force following the introduction of the third roaming regulation in 2014 (€0.05 per minute for calls made, €0.02 per SMS sent and €0.05 per MB of data used (excl. VAT). The EU also agreed new legislation covering roaming under the Connected Continent Regulation on 30 June 2015. Under this regulation roaming must be abolished by 15 June 2017 within the 28 EU member states.

However, a fair use policy will be introduced to prevent abusive or anomalous usage of regulated retail roaming services by roaming customers. An example scenario covered by the fair use policy would be if a customer permanently stayed abroad while using a domestic subscription for his home country. In addition, the European Commission will undertake a review on the wholesale roaming market by 15 June 2016.

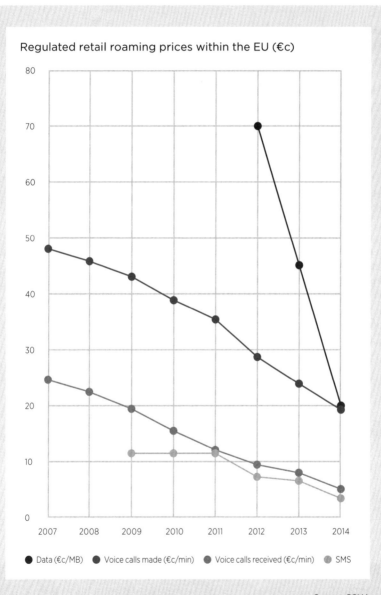

Regulated retail roaming prices within the EU (€c)

Legend:
● Data (€c/MB) ● Voice calls made (€c/min) ● Voice calls received (€c/min) ● SMS

Source: GSMA

Mobile Termination Rates

Background

Mobile termination rates (MTRs) refer to the fees charged by operators to connect a phone call that originates from a different network.

The setting of regulated MTRs continues to be the focus of regulatory attention in both developed and developing countries, and many different approaches have been developed for the calculation of appropriate termination charges.

Regulators have generally concluded that the provision of call termination services on an individual mobile network is, in effect, a monopoly. Therefore, with each operator enjoying significant market power, regulators have developed various regulations, most notably the requirement to set cost-oriented prices for call termination.

Debate

How should the appropriate, regulated rate for call termination be calculated?

Is the drive towards ever-lower mobile termination rates, especially in Europe, a productive and appropriate activity for regulators?

Once termination rates have fallen below a certain threshold, is continued regulation productive?

What is the long-term role of regulated termination rates in an all-IP environment?

Intervening in a competitive market is far more complex and challenging than the traditional utility regulation of the kind normally applied to monopolies in gas, electricity and fixed-line telecommunications. With mobile, every action is more finely calibrated. The benefits of intervention are more ambiguous and the error costs larger.

— Stewart White, former Group Public Policy Director, Vodafone

Industry Position

Regulated mobile termination rates should accurately reflect the costs of providing termination services.

Beyond a certain point, evidence suggests that a focus on continued reductions in MTRs is not beneficial.

The setting of regulated MTRs is complex and requires a detailed cost analysis as well as a careful consideration of its impact on consumer prices and, more broadly, on competition.

MTRs are wholesale rates, regulated in many countries, where a schedule of annual rate changes has been established and factored into mobile network operators' business models. Unsignaled, unanticipated alterations to these rates have a negative impact on investor confidence.

We believe the setting of MTRs is best done at a national level, where local market differences can be properly reflected in the cost analysis, therefore extraterritorial intervention is not appropriate.

Resources:
Report: The Impact of Recent Cuts in Mobile Termination Rates Across Europe
Report: The Setting of Mobile Termination Rates
Report: Comparison of Fixed and Mobile Cost Structures
Report: Regulating Mobile Call Termination, Vodafone

Case Study

Impact of Accelerated MTR Reductions in Europe

In 2009, the European Commission recommended an accelerated reduction in mobile termination rates, proposing that Member States implement rates based on the pure Long Run Incremental Cost (LRIC). It reasoned that the MTR cuts would reduce mobile prices and therefore increase usage, while also helping smaller mobile network operators to be price-competitive.

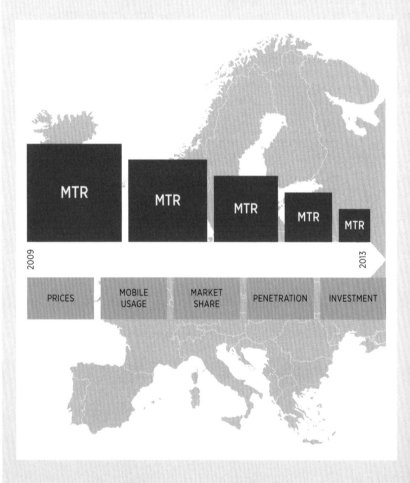

Frontier Economics was commissioned in 2012 by Vodafone to determine whether the policy — to the extent that it has been applied in EU countries — has had the intended effect. Among the findings are these points:

1. There is no evidence that faster MTR cuts have led to lower mobile prices.

Although mobile prices in Europe have been falling, there is no support for the view that this has been driven by MTR cuts.

2. There is no evidence that MTR cuts are increasing usage.

Since 2009, usage has not increased at an accelerated rate, and countries with the largest MTR cuts have not had the largest increases in usage.

3. There is limited evidence of any link between MTR reductions and the market share of smaller operators.

While nearly all of the smallest operators experienced an increase in their market share, no link with the MTR reductions was observed.

4. Accelerated MTR cuts could be detrimental to network investment and mobile penetration.

While it is too early to conclude whether the MTR cuts are having a detrimental effect, there is some indication that mobile penetration and investment are being adversely affected.

Source: Frontier Economics, 'The Impact of Recent Cuts in Mobile Termination Rates Across Europe', May 2012

Net Neutrality

Background

In 1973, work began on establishing a global network of networks, an 'internetworking' project that became the internet. The objective was to design a network that was self-sustaining, and that would be able to run applications not yet designed. The solution was simple and rested on two rules: there can be no central control, and the network cannot be optimised for any single application.

Today's net neutrality debate has evolved from these two rules. Networks that were connected to the internet had to communicate via common protocols, primarily the Transition Control Protocols and the Internet Protocol (TCP/IP), an architecture that rendered network performance as best efforts and assumed the intelligence would be either in applications or at the user interface (i.e., on computer terminals).

While there is no single definition of net neutrality, it is often used to refer to issues concerning the optimisation of traffic over networks. Some argue that it is necessary to legislate that all traffic carried over a network be treated in the same way. Others advocate that flexibility to offer varying service levels, for different applications, enhances the user experience.

Mobile operators face unique operational and technical challenges in providing fast, reliable internet access to their customers, due to the shared use of network resources and the limited availability of spectrum.

Unlike fixed broadband networks, where a known number of subscribers share capacity in a given area, the capacity demand at any given cell site is much more variable, as the number and mix of subscribers constantly changes, often unpredictably. The available bandwidth can also fluctuate due to variations in radio frequency signal strength and quality, which can be affected by weather, traffic, speed and the presence of interfering devices such as wireless microphones.

Not all traffic makes equal demands of a network; for example, voice traffic is time-sensitive while video streaming typically requires large amounts of bandwidth. Networks need to be able to apply network management techniques to ensure each traffic type is accommodated.

Just as content providers offer differentiated services such as standard and premium content for different prices, mobile network operators will offer different bandwidth products to meet different consumer needs. Customers are benefitting from these tailored solutions; only those who want to use premium services will have to pay for the associated costs.

— GSMA

Debate

Should networks be able to manage traffic and prioritise one traffic type or application over another?

For mobile networks, which have finite capacity, should fixed-line rules apply?

In some cases, net neutrality rules are being considered in anticipation of a problem that has yet to materialise. Is this an appropriate approach to regulation?

Industry Position

To meet the varying needs of consumers, mobile network operators need the ability to actively manage network traffic.

It is important to maintain an open internet. To ensure it remains open and functional, mobile operators need the flexibility to differentiate between different types of traffic. However, within the context of a single traffic type, operators should not discriminate in favour of any one content provider.

Regulation that affects network operators' handling of mobile traffic is not required. Any regulation that limits their flexibility to manage the end-to-end quality of service and provide consumers with a satisfactory experience is inherently counterproductive.

In considering the issue, regulators should recognise the differences between fixed and mobile networks, including technology differences and the impact of radio frequency characteristics.

Consumers should have the ability to choose between competing service providers on the basis of being able to compare performance differences in a transparent way.

Mobile operators compete along many dimensions, such as pricing of service packages and devices, different calling and data plans, innovative applications and features, and network quality and coverage. The high degree of competition in the mobile market provides ample incentives to ensure customers enjoy the benefits of an open internet.

Resources:

Net Neutrality on the GSMA website

FCC Filing: GSMA Comments on the Open Internet Proceeding, 15 July 2014

Deeper Dive

Traffic Management Is an Efficient and Necessary Tool

Traffic growth, the deployment of next-generation technologies and the emergence of new types of services are presenting mobile network operators with a huge challenge: how to manage different types of traffic over a shared network pipe, while providing subscribers with a satisfactory quality of service that takes into account different consumer needs and service attributes.

With finite capacity, mobile networks experience congestion. Mobile operators use traffic management techniques to efficiently manage network resources, including spectrum, and to support multiple users and services on their networks. Congestion management is essential to prevent the network from failing during traffic peaks, and to ensure access to essential services.

Traffic management techniques are applied at different layers of the network, including admission control, packet scheduling and load management. In addition, operators need to cater to different consumer preferences, so customers can access the services they demand. Traffic management is therefore an efficient and necessary tool for operators to manage the flow of traffic over their network and provide fair outcomes for all consumers.

Mobile operators need the flexibility to experiment and establish new business models that align investment incentives with technological and market developments, creating additional value for their customers. As the operational and business models of networks evolve, a whole host of innovative services and business opportunities will emerge.

The current competitive market is delivering end-user choice, innovation and value for money for consumers and no further regulatory intervention related to provision of IP-based services is necessary. The commercial, operational and technological environment in which these services are offered is continuing to develop, and any intervention is likely to impact the development of these services in a competitive context.

Traffic management techniques are necessary and appropriate in a variety of operational and commercial circumstances:

Network integrity

Protecting the network and customers from external threats, such as malware and denial-of-service attacks.

Child protection

Applying content filters that limit access to age-appropriate content.

Subscription-triggered services

Taking the appropriate action when a customer exceeds the contractual data-usage allowance, or offering charging models that allow customers to choose the service or application they want.

Emergency calls

Routing emergency call services.

Delivery requirements

Prioritising real-time services, such as voice calls, as well as taking into account the time sensitivities of services such as remote alarm monitoring.

Over-the-Top Voice and Messaging Communications Apps

Background

The combination of mobile broadband access, smartphones and internet technology has led to the emergence of a new breed of consumer mobile voice and messaging communication services provided by internet-based companies, often referred to as over-the-top service providers (OTTs). These services are providing consumers with additional choices in how they communicate with each other. According to industry research, global instant messaging volumes from OTT providers already exceed SMS volumes. Research also shows that voice-over-IP (VoIP) now accounts for over 40% of international voice traffic . Fuelling this trend, OTTs are increasingly developing techniques to influence users' decisions about whether calls and messages should go through the Public Switched Telephone Network (PSTN) or the internet.

OTT communications services are typically offered in competition with, and as direct substitutes to, the circuit-switched voice and SMS services provided by mobile operators, but they are typically not properly considered in the market analysis carried out by regulators. Due to the global nature of the internet, and because they have not been considered as equivalent to traditional communication services, many OTT communications services are able to sit outside the scope of sector specific national or regional regulatory and fiscal obligations (e.g., data privacy, legal interception, emergency calls, universal service contribution, national specific taxes, consumer rights and quality of service) that have been put in place to protect consumers and ensure that all providers make a fair and proportionate contribution to local economic growth through investment, employment and tax.

As OTT communications services become more and more popular, they increasingly render a number of regulations designed to address alleged network bottlenecks, such as termination and roaming, unjustified.

Debate

Should OTT services be subject to the same regulatory obligations that apply to calls and messages carried over the PSTN?

Does the fact that OTT players currently sit outside the scope of sector-specific regulations provide them with a competitive advantage over traditional telecoms providers?

Everybody knows today that with telecom service providers and OTT [players], there are unbalanced relations and we have to find a better balance.

— European Commission Vice President Andrus Ansip, March 2015

Industry Position

The mobile industry supports and promotes fair competition as the best way to stimulate innovation and investment for the benefit of consumers and to spur economic growth, and believes both objectives will be best served by the principle of 'Same Rules for the Same Service'. The growth in competition between different types of service provider calls for a move towards shared rules that are lighter touch than those applicable in less competitive environments.

The principle of 'Same Rules for the Same Service' maintains that where regulation is considered to be necessary, all equivalent consumer voice and messaging services should be subject to the same regulatory and fiscal obligations, regardless of the underlying technology, geographic origin or whether they are delivered by a mobile operator or OTT service provider. This will help to improve consumer confidence and trust in using internet-based services by ensuring a consistent approach to issues such as transparency, quality of service and data privacy. Consistent application of regulatory obligations will also support legitimate law enforcement and national security activities.

While the same rules should apply to the same services, these are not necessarily the rules that apply today to telecommunications services. There is a need for a forward-looking regulatory framework for communications services that is fit for purpose for a digital world. This framework must be driven by clear policy requirements around consumer protection, innovation, investment and competition.

By adopting a policy framework built around 'Same Rules for the Same Service', and properly recognising the competitive constraint imposed on mobile network operators by the fact OTTs currently play to different rules, national governments and regulators will be enabling an environment of fair and sustainable competition that promotes the best interests of consumers and fosters economic growth.

Resources:
The TeleGeography Report 2014
Deloitte TMT Predictions 2014

Passive Infrastructure Providers

Background

Many mobile network operators share infrastructure on commercial terms to reduce costs, avoid unnecessary duplication and to expand coverage cost-effectively in rural areas.

The most commonly shared infrastructure is passive infrastructure, which may include: land, rights of way, ducts, trenches, towers, masts, dark fibre and power supplies, all of which support the active network components required for transmission and reception of signals.

Infrastructure sharing is arranged through bilateral agreements between mobile network operators to share the specific towers, strategic sharing alliances, the formation of joint infrastructure companies between mobile operators or via independent companies providing towers and other passive infrastructure.

Increasingly, independent tower companies provide tower-sharing facilities to network operators. Several countries have established regulatory frameworks based on registration that encourage passive infrastructure sharing arrangements and provide regulatory clarity for network operators and independent passive infrastructure providers. While regulatory authorities in almost all countries are supportive of passive infrastructure sharing arrangements, a lack of regulatory clarity exists in some countries, particularly in relation to independent tower companies.

Debate

What benefits do independent tower companies offer to mobile operators?

Should passive infrastructure sharing ever be mandated by the regulatory authority?

What steps should regulators take to provide clarity to tower companies and mobile operators?

Industry Position

Licensed network operators should be able to share passive infrastructure with other licensed network operators and outsource passive infrastructure supply to passive infrastructure providers without seeking regulatory approval.

Sharing passive infrastructure on commercial terms enables operators to reduce capital and operating expenditure without affecting investment incentives or their ability to differentiate and innovate.

Infrastructure sharing provides a basis for industry to expand coverage cost-effectively and rapidly, while retaining competitive incentives. Regulation of passive infrastructure sharing should be permissive, but should not mandate such arrangements.

In markets with licensing frameworks that do not already provide for the operation of independent tower companies, regulatory authorities (or the responsible government department) should either permit independent passive infrastructure companies to operate without sector-specific authorisation or establish a registration scheme for such companies. The scheme should be a simple authorisation that provides for oversight of planning-related matters while making a clear distinction with the licensing framework applicable to electronic communications network and service providers.

Registered providers should be permitted to construct and acquire passive infrastructure that is open to sharing with network operators, provide (e.g., sell or lease) passive infrastructure elements to licensed operators, and supply ancillary services and facilities essential to the provision of passive infrastructure.

Mobile network operators should be permitted to make use of infrastructure from passive infrastructure companies through commercial agreements without explicit regulatory approval. Infrastructure sharing agreements should be governed under commercial law and, as such, be subject to assessment under general competition law.

Public authorities should provide licensed operators and passive infrastructure providers with access to public property and rights of way on reasonable terms and conditions. Governments, seeking to support national infrastructure development, should ensure swift approval for building passive infrastructure, and environmental restrictions should reflect globally accepted standards.

Taxation and fees imposed on independent tower or passive infrastructure companies should not act as a barrier to the evolution of this industry, which makes possible more efficient, lower-cost forms of infrastructure supply.

Resources:
AT Kearney: The Rise of the Tower Business
Financial Times: Bharti Airtel to Sell 3,100 Telecom Towers

Quality of Service

Background

The quality of a mobile data service is characterised by a small number of important parameters, notably speed, packet loss, delay and jitter. It is affected by factors such as mobile signal strength, network load, and user device and application design.

Mobile network operators must manage changing traffic patterns and congestion, and these normal fluctuations result in customers experiencing a varying quality of service.

Connection throughput is seen by some regulatory authorities as an important attribute of service quality. However, it is also the most difficult to define and communicate to mobile service users. Mobile throughput can vary dramatically over time, and throughput is not the only product attribute that influences consumer choice.

Debate

Is it necessary for regulators to set specific targets for network quality of service in competitive markets?

Is it possible to guarantee minimum quality levels in mobile networks, which vary over time according to the volume of traffic being carried and the specific, local signal-propagation conditions?

Which regulatory approach will protect the interests of mobile service customers while not distorting the market?

Industry Position

Competitive markets with minimal regulatory intervention are best able to deliver the quality of mobile service customers expect. Regulation that sets a minimum quality of service is disproportionate and unnecessary.

The quality of service experienced by mobile consumers is affected by many factors, not all of which are under the control of operators. Defining specific quality targets is neither proportional nor practical.

Some of the factors affecting the quality of service are beyond the control of operators, such as the device type, application and propagation environment.

Mobile networks are technically different from fixed networks; they make use of shared resources to a greater extent and are more traffic-sensitive.

Mobile operators need to deal with continually changing traffic patterns and congestion, within the limits imposed by finite network capacity, where one user's traffic can have a significant effect on overall network performance.

The commercial, operational and technological environment in which mobile services are offered is continuing to develop. Mobile operators must have the freedom to manage and prioritise traffic on their networks. Regulation which rigidly defines a particular service quality level is unnecessary and is likely to impact the development of these services.

Competitive markets with differentiated commercial offers and information that allows consumers to make an informed choice deliver the best outcomes. If regulatory authorities are concerned about quality of service, they should engage in dialogue with the industry to find solutions that strike the right balance on transparency of quality of service.

Resources:
GSMA Latin America: QoS
GSMA Response to the EC Consultation on Traffic Management, Transparency and Switching

Deeper Dive

A Network of Interconnections

Offering a dependable quality of service is a priority for mobile network operators, as it allows them to differentiate the internet access service they provide from that of their competitors and meet customer expectations. However, mobile operators have little control over many of the parameters that can affect their subscribers' experience.

Factors beyond operators' control include:

The type of device and application being used.

The changing usage patterns in a mobile network cell at different times of day.

The movements and activities of mobile users, such as travel, events or accidents.

Obstacles and distance between the terminal and antennas.

The weather, especially rain.

In addition, the quality of internet access that users experience depends on the quality provided by each of the data paths followed. The ISP only has control of the quality of service in its section of the network.

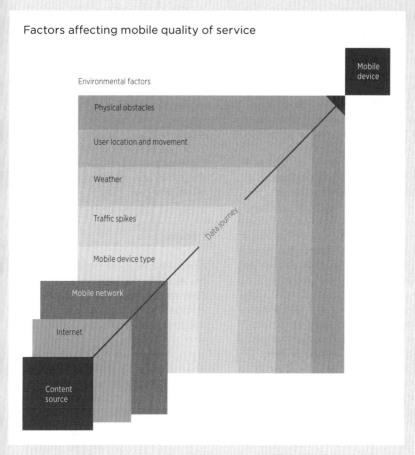

Factors affecting mobile quality of service

Environmental factors
- Physical obstacles
- User location and movement
- Weather
- Traffic spikes
- Mobile device type

Mobile device

Mobile network

Internet

Content source

Data journey

For these reasons, regulation concerning the quality of mobile internet service can be counterproductive. Regulation that does not consider the nature of mobile networks and the competitive workings of these services can be an obstacle to their development, widening the digital divide and promoting an inefficient use of the capital invested in networks.

Single Wholesale Networks

Background

Policymakers in some countries are considering establishing a single wholesale network (SWN) instead of relying on competing mobile networks to deliver mobile broadband services in their country. Most of these proposals specify at least partial network ownership and financing by the government.

While there are variations in the SWN proposals discussed by different governments, SWNs can be generally defined as government-initiated network monopolies that compel mobile operators and others to rely on wholesale services provided by the SWN as they serve and compete for retail customers.

SWNs would represent a radical departure from the approach to mobile service provision that has been favoured by policymakers for the past 30 years — namely, to license a limited number of competing mobile network operators, which are usually under private ownership.

In 2000, there were almost as many countries served by a single mobile network as by competing networks. Only 30 countries today, representing less than three per cent of the world's population, are served by a single mobile network. Since 2000, network competition has produced unprecedented growth and innovation in mobile services, particularly in developing countries. For example, the number of unique mobile subscribers has almost tripled in developed countries from 339 million in 2000 to over 880 million today, while in developing countries the number of subscribers has increased from 131 million to more than 2.5 billion.[1]

Supporters of SWNs argue that they can address some issues better than the traditional model of network competition in some markets. These concerns generally include inadequate or slow coverage in rural areas, inefficient use of radio spectrum and concerns that the private sector may lack incentives to maximise coverage or investment.

Debate

Are SWNs likely to increase the quality and reach of next-generation mobile broadband, compared with the existing approach of network competition?

What alternative policies should be considered before adopting a monopoly wholesale network model?

Industry Position

SWNs will lead to worse outcomes for consumers than network competition.

Some SWN supporters claim that SWNs will deliver greater network coverage than network competition can, but this claim often reflects the existence of public subsidies and other forms of support for the SWN, which are not available to competing network operators. The claim is therefore unsupported. Network competition can deliver coverage in areas where duplicate networks are uneconomic through voluntary network sharing and the commercial incentive of being first to market in a particular area.

The benefits of network competition go beyond coverage. Innovation is a key driver of consumer value at the national level, and this occurs in networks as well as services and devices. While mobile technologies are typically developed at the international level, the speed at which they become available to consumers depends on national policies and market structures. In practice, single networks have been much slower to expand coverage, perform upgrades and to embrace new technologies such as 3G, and SWNs can be expected to prompt less innovation than network competition.

To achieve the objectives of their proponents, SWNs would need to evolve into regulated monopolies, leading to worse long-term outcomes for consumers. As monopolies, SWNs will always have incentives to keep prices high and reduce expenditures, including network deployment to increase coverage. Although regulation can attempt to ensure SWNs mimic the outcomes of a competitive market, it will not fully succeed.

SWNs may co-exist for some period with existing networks. As SWNs are likely to be supported by governments, this will likely lead to a distortion of competition. Co-existence is also likely to increase uncertainty, which will have a dampening effect on investment in mobile broadband services.

The evidence suggests that the design, financing and implementation of SWNs are likely to prove challenging and that there is a significant risk of failure.

Although a publically funded SWN could deliver coverage in areas where privately funded competing networks would not be willing to expand into, the correct approach is to consider how public subsidies could be used to extend the benefits of network competition to those areas. This can be achieved in a variety of ways, including coverage obligations and other forms of subsidy, such as the award of contracts to cover particular areas using public funds.

[1]   Source: GSMAi

Resources:
Report: Assessing the Case for Single Wholesale Networks in Mobile Communications,
Frontier Economics, August 2014

Taxation

Background

The mobile telecommunications sector has a positive impact on economic and social development, creating jobs, increasing productivity and improving the lives of citizens.

Sector-specific taxes are levied on mobile consumers and operators in many countries. These include special communication taxes, such as excise duties on mobile handsets and airtime usage, and revenue-share levies on mobile operators. These taxes contribute to a high tax burden on the mobile sector that exceeds the burden on other sectors.

Some countries have applied a surtax on international inbound call termination (SIIT), which can have the effect of increasing international call prices and act as a tax on other countries' citizens.

Debate

Do sector-specific taxes deliver short-term government income at the expense of a country's long-term additional tax revenues resulting from increased economic growth?

Industry Position

Governments should reduce or remove mobile-specific taxes because the resulting social impact and long-term positive impact on gross domestic product, and hence tax revenues, will outweigh any short-term contributions to governments' budgets.

Taxes should align with internationally recognised principles of effective tax systems. In particular:

- Taxes should be broad-based — different taxes have different economic properties and, in general, broad-based consumption taxes are less distortionary than taxation on income or profits.

- Taxes should account for sector and product externalities.

- The tax and regulatory system should be simple, easily understandable and enforceable.

- Dynamic incentives for the operators should be unaffected — taxation should not disincentivise efficient investment or competition in the ICT sector.

- Taxes should be equitable and the burden of taxation should not fall disproportionately on the lower income members of society.

Analytical research has demonstrated that although the telecommunication / ICT sector tax revenues play an important role in supporting national public services, this role must be weighed against the potentially adverse effects that taxation can bring to the growth of the telecommunication / ICT sector, broadband penetration, and national economic growth.

— Brahima Sanou, Director of the Telecommunication Development Bureau at the International Telecommunication Union (ITU)

Discriminatory, sector-specific taxes deter the take-up of mobile services and can slow the adoption of information and communication technology (ICT). Lowering such taxes benefits consumers, businesses and socio-economic development.

Governments often levy special taxes to finance spending in sectors where private investment is lacking; however, this approach is inefficient. Fiscal policy that applies a special tax to the telecommunications sector causes distortions that deter private spending and, in the end, diminish welfare by preventing the realisation of the positive spillovers that mobile provides throughout the economy.

Emerging economies need to align their approach to taxing mobile broadband with national ICT objectives. If broadband connectivity is a key social and economic objective, taxes must not create an obstacle to investment in broadband networks or adoption and usage of mobile broadband by consumers. Lowering the taxation burden on the sector increases mobile take-up and use, creating a multiplier effect in the wider economy. Taxing international calls negatively impacts consumers, businesses and citizens abroad, damaging a country's competitiveness.

The Impact of Tax Rebalancing on the Economy

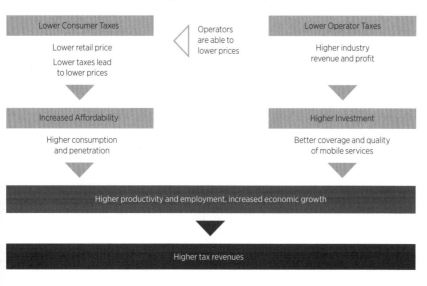

Resources:

GSMA Mobile Taxation Research and Resources
Digital Inclusion and Mobile Sector Taxation 2015

Taxes and Fees on Mobile Consumers and Operators

Mobile operators have repeatedly raised concerns that their customers are facing an undue burden from taxation, compared to other goods. The taxation and fees burden on the mobile sector consists of a wide range of charges. On the consumer side, this includes taxes on handset purchases and connection activation, as well as calls, messages and data access.

In addition to these consumer-facing charges, mobile operators also face a range of other charges including licensing fees, corporation tax, revenue charges and many more. The extent to which these charges fall on operators or consumers depends on individual market conditions. Some taxes may be absorbed by operators in the form of lower profits, while others may be passed through to consumers as higher prices or a combination of the two.

Research by Deloitte for the GSMA revealed that:

• Across 26 selected countries, the total tax and fee payments from the mobile sector amounted to $39 billion in 2013, while market revenues were $117.5 billion.

• Total mobile tax payments from taxation on both consumers and operators are estimated to range from 10.6 per cent as a proportion of market revenues in Nigeria to 58.3 per cent in Turkey, excluding non-recurring payments such as spectrum auction fees.

• Over the sample, sector-specific taxes make up on average 32.1 per cent of the recurring payments on mobile services, including taxes on both consumers and operators.

• Taxes and fees on each mobile connection cost $35.6 on average per year across 26 selected countries.

Moreover, the gap between telecoms and other sectors appears to be growing over the same period. The burden on mobile services has increased at an average of 2.1 per cent per year, yet the overall tax burden in the countries considered as a percentage of gross domestic product (GDP) has on average declined at an annual rate of -0.2 per cent.

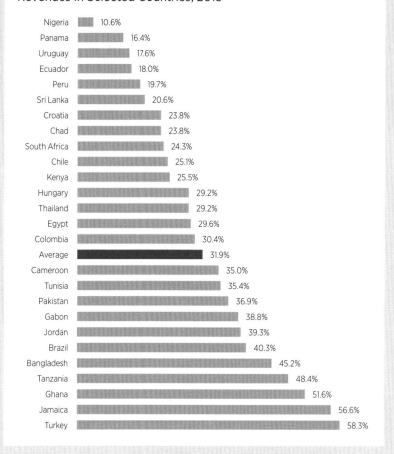

Total Tax and Fee Payments as a Proportion of Mobile Revenues in Selected Countries, 2013

Country	Percentage
Nigeria	10.6%
Panama	16.4%
Uruguay	17.6%
Ecuador	18.0%
Peru	19.7%
Sri Lanka	20.6%
Croatia	23.8%
Chad	23.8%
South Africa	24.3%
Chile	25.1%
Kenya	25.5%
Hungary	29.2%
Thailand	29.2%
Egypt	29.6%
Colombia	30.4%
Average	31.9%
Cameroon	35.0%
Tunisia	35.4%
Pakistan	36.9%
Gabon	38.8%
Jordan	39.3%
Brazil	40.3%
Bangladesh	45.2%
Tanzania	48.4%
Ghana	51.6%
Jamaica	56.6%
Turkey	58.3%

Source: Deloitte analysis based on GSMA Intelligence database and operator data

The cost of taxation on mobile consumers

The costs borne by consumers in order to own and use a mobile phone can be expressed as the Total Cost of Mobile Ownership (TCMO) and includes expenditure on calls, SMS and data, as well as connection/activation and handset costs. Research by GSMA and Deloitte finds that taxes applied directly on mobile consumers represented 20 per cent of TCMO across 110 countries in 2014. Notably:

- Today, among the 110 countries surveyed, 44 levy taxes that are specific to or are applied at higher rates on mobile services. Of these 44 countries, 17 are in Africa, seven are in Latin America, seven are in Asia Pacific and five are in Middle East and North Africa (MENA).

- 37 countries levy industry-specific taxes on mobile usage, such as airtime and data tax or additional VAT over the standard rate. 24 countries apply specific usage taxes on mobile data, and seven impose higher VAT on these services compared to the standard rate.

- 25 countries impose a special tax or additional VAT on handsets, in addition to custom duties on imported devices that are in some cases higher for mobile than for other goods.

- 10 countries, including Jamaica, Tunisia, Pakistan and Bangladesh, apply an activation tax that is paid upon purchase or activation of a SIM card, and hence represents a barrier to access for lower-income consumers.

- Compared to goods or services that are only subject to VAT, mobile services receive about 33 per cent more taxation in the countries that impose mobile-specific consumer taxes. In these countries, consumers pay on average $8 more in tax than on a standard good for each $100 spent.

Consumer Taxes as a Proportion of TCMO, 2014

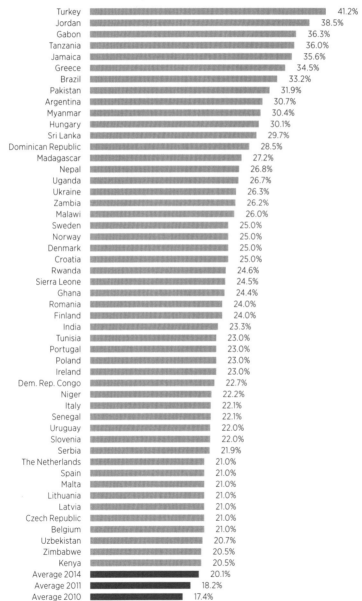

Country	Percentage
Turkey	41.2%
Jordan	38.5%
Gabon	36.3%
Tanzania	36.0%
Jamaica	35.6%
Greece	34.5%
Brazil	33.2%
Pakistan	31.9%
Argentina	30.7%
Myanmar	30.4%
Hungary	30.1%
Sri Lanka	29.7%
Dominican Republic	28.5%
Madagascar	27.2%
Nepal	26.8%
Uganda	26.7%
Ukraine	26.3%
Zambia	26.2%
Malawi	26.0%
Sweden	25.0%
Norway	25.0%
Denmark	25.0%
Croatia	25.0%
Rwanda	24.6%
Sierra Leone	24.5%
Ghana	24.4%
Romania	24.0%
Finland	24.0%
India	23.3%
Tunisia	23.0%
Portugal	23.0%
Poland	23.0%
Ireland	23.0%
Dem. Rep. Congo	22.7%
Niger	22.2%
Italy	22.1%
Senegal	22.1%
Uruguay	22.0%
Slovenia	22.0%
Serbia	21.9%
The Netherlands	21.0%
Spain	21.0%
Malta	21.0%
Lithuania	21.0%
Latvia	21.0%
Czech Republic	21.0%
Belgium	21.0%
Uzbekistan	20.7%
Zimbabwe	20.5%
Kenya	20.5%
Average 2014	20.1%
Average 2011	18.2%
Average 2010	17.4%

Source: Deloitte analysis based on GSMA Intelligence database and operator data

Universal Service Funds

Background

Universal service — characterised by telecommunications service that is available, accessible and affordable — is a policy goal of many governments.

Some countries have established universal service funds (USFs) on the premise that operators are unable to extend service to some areas without financial support.

USFs are typically funded by levies on telecommunication sector revenues.

In these cases, operators continue to be required to contribute a share, despite the expansion of service to the vast majority of countries' citizens and increasingly large accumulations of undisbursed funds.

According to a 2013 report commissioned by the GSMA, fewer than one-eighth of the 64 USFs studied are achieving their targets, and more than one-third have yet to disburse any of the funds they have collected. Nevertheless, the levies continue to be required from the sector.

Debate

Are USFs an effective way to extend voice and data connectivity to underserved citizens?

What alternative strategies could be more effective?

How relevant are USFs in mature markets?

Industry Position

Governments should phase out USFs and discontinue collecting USF levies. Existing USF monies should be returned to operators and used to extend mobile services to remote areas.

Liberalised markets and private-sector investment have delivered telecommunication services to the majority of the world's population, a trend that the industry considers will continue.

Few USFs have successfully expanded access to telecommunication services, as is their objective, yet they continue to accumulate large sums of money.

There is little evidence that USFs are an effective way to achieve universal service goals and many have, in fact, been counterproductive, because they tax communications customers, including in rural areas, and therefore raise the barrier to rural investment.

USFs that already exist should be targeted, time-bound and managed transparently. The funds should be allocated in a competitive and technically neutral way, in consultation with the industry.

Governments should consider incentives that facilitate market-based solutions. They can help by removing sector-specific taxes, stimulating demand and developing the supporting infrastructure. Alternative solutions such as public-private partnerships should be explored in preference to USFs for the extension of communications to rural and remote areas.

Resources:
Report: Survey of Universal Service Funds, Key Findings

Estimated USF Funds Available

Despite the admirable goals that led to the creation of USFs during the early stages of telecoms liberalisation, there is now considerable doubt about their practicality and efficacy. A large proportion of USF monies collected remain undisbursed, and the structure of many USFs is too rigid to respond to rapid technological changes and societal requirements.

Africa

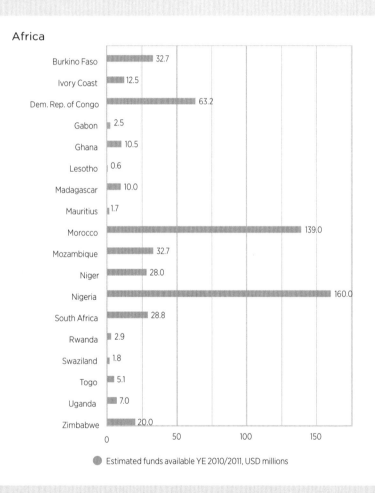

Estimated funds available YE 2010/2011, USD millions

Source: GSMA, Survey of Universal Service Funds, April 2013

Asia Pacific

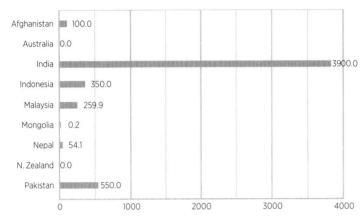

● Estimated funds available YE 2010/2011, USD millions

Source: GSMA, Survey of Universal Service Funds, April 2013

America

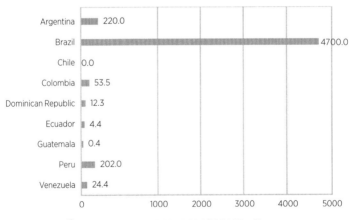

● Estimated funds available YE 2010/2011, USD millions

Source: GSMA, Survey of Universal Service Funds, April 2013

Spectrum Management and Licensing

Data traffic on mobile networks is skyrocketing as consumers and businesses use smartphones, tablets, laptops and other devices to access email and entertainment, mapping and messaging, browsing and banking, as well as social networking and sharing services. As the nascent Internet of Things industry grows, it too will place significant extra demand on mobile data services.

To meet this explosion in demand, mobile operators need more spectrum. Sufficient, internationally harmonised spectrum is essential to ensuring the quality of service that consumers and businesses have come to expect, and rely on, from mobile networks.

The GSMA is very active at the national, regional and global levels to advocate for the timely identification and release of more spectrum for mobile broadband. In this regard, we work with national governments and regulators, with regional organisations and the International Telecommunication Union (ITU).

The GSMA also serves as a clearinghouse for sector research and market data. Because spectrum management has many facets — including issues such as interference, spectrum auctions and licence processes — the GSMA contributes on behalf of mobile operators to the work of regulators with market projections, analysis, regulatory guidance and policy recommendations based on objective data and recognised best practice. Many of these reports are referenced in this handbook.

2.1GHz Frequency Band

Background

Paired spectrum refers to mobile frequency bands, such as the 2.1GHz band, with separate allocations for uplink and downlink.

The 2.1GHz band, referring to 1.7/2.1GHz (3GPP band 4: 1710–1755MHz paired with 2110–2155MHz) in most countries in the Americas, and 1.9/2.1GHz (3GPP band 1: 1920–1980MHz paired with 2110–2170MHz) elsewhere, has been licensed for 3G mobile services in most markets. However, several countries are yet to release this spectrum for mobile.

Excessive per-MHz spectrum costs are an issue in certain markets, as a result of governments seeking to ration spectrum in order to maximise short-term revenue from the auctions.

Debate

Is there any reason regulators should not have already licensed the entire 2.1GHz band to mobile operators?

How should the licences be awarded to maximise value to society?

Industry Position

The 2.1GHz frequency band should be released in all markets for mobile broadband services, preferably in blocks larger than 2x10MHz per operator.

Releasing the 2.1GHz band for mobile is critical for governments to enable the digital economy and to prevent a growing digital divide.

In certain markets, due to political instability or regulatory uncertainty, investors (including mobile network operators) may not advocate immediate licence allocation; in these cases the optimal timing of spectrum allocation depends on local factors.

Governments should not look to generate excessive fees from the licensing of 2.1GHz spectrum, as this will artificially limit demand, negatively impact network deployment, increase consumer prices and limit the economic benefits. Excessive fees can also result in unsold spectrum, further impeding policy goals of delivering mobile broadband access to everyone.

Resources:
Report: Licensing to Support the Broadband Revolution
GSMA Europe response to the public consultation on the introduction of harmonised technical conditions for the terrestrial 2GHz band
Report: Momentum Building in the AWS Band (GVP)

2.6GHz Frequency Band

Background

The International Telecommunication Union (ITU) has identified the 2.6GHz band (2500–2690MHz) as a global allocation for mobile telecommunications. The 2.6GHz radio spectrum band is a 'capacity band' for mobile broadband, well suited for the next generation of mobile technologies that respond to the soaring demand for data-heavy content, such as video. The band is identified for mobile in all regions and has the potential to be used in a harmonised manner on a global basis. The harmonised use will result in economies of scale for industry and cheaper handsets for consumers, as well as increased flexibility for roaming.

The ITU has proposed several possible band plans, including:

- Option 1: 2x70MHz for FDD with a 50MHz TDD in the centre gap.

- Option 2: FDD only.

- Option 3: Flexible TDD/FDD arrangement.

Excessive per-MHz spectrum costs are an issue in certain markets, as a result of governments seeking to ration spectrum in order to maximise short-term revenue from the auctions.

2.6GHz Band Plan – Option 1

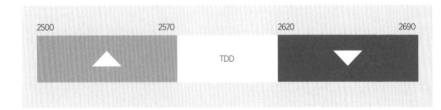

Debate

Should the 2.6GHz band be released in conjunction with the Digital Dividend band (700MHz/800MHz) to meet urban and rural coverage and capacity needs for mobile broadband?

Which band plan option is best?

Industry Position

We support ITU Option 1 for a globally harmonised 2.6GHz capacity band. Global momentum for the 2.6GHz band is behind ITU Option 1, with countries such as Brazil, Canada, Chile, Qatar, UAE, Russia and the UK having already assigned the spectrum to mobile operators under this band plan. Where auctions have offered flexibility, markets have chosen standard band arrangements. The 2.6GHz band will be critical in meeting the capacity requirements of mobile broadband.

ITU Option 1 is a technology-neutral option, supporting both TDD and FDD technologies (e.g., LTE and Wi-MAX). The spectrum available in the 2.6GHz band provides for large carriers such as 2x20MHz, which is ideal for the deployment of LTE:

- To improve network performance, offering faster data transmission and greater capacity.

- To reduce deployment costs.

- To improve handset performance.

Higher frequencies (e.g., 2.6GHz) are better suited to high data rates required to serve large numbers of users in urban areas, airports and other high-traffic areas. Governments should not look to generate excessive fees from the licensing of 2.6GHz spectrum, as this will artificially limit demand, negatively impact network deployment, increase consumer prices and limit the potential economic benefits. Excessive fees also can impede policy goals of delivering mobile broadband access to everyone.

Resources:

Brochure: The 2.6GHz Spectrum Band: An Opportunity for Global Mobile Broadband

Report: Taiwan — Economic Impact of Wireless Broadband

Report: The Socio-Economic Benefit of Allocating Harmonised Spectrum in the Kingdom of Saudi Arabia

Report: The Benefits of Releasing Spectrum for Mobile Broadband in Sub-Saharan Africa

Report: Arab States Mobile Observatory 2013

Deeper Dive

Band Characteristics – Capacity vs. Coverage

Not all radio frequencies are equal, and mobile network operators require access to a range of frequency bands to cost effectively offer a high-quality service for different locations with different population densities and different demands on the network.

In general, lower-frequency signals reach further beyond the visible horizon, and are better at penetrating rain or buildings. These lower radio frequencies are sometimes called coverage bands because, as a rule, an operator can serve a larger area with one base station.

The capacity of a wireless connection for data or voice calls is dependent on the amount of spectrum it uses — the channel bandwidth — and wider channel bandwidths are more readily available at higher frequencies. For many wireless applications, the best trade-off of these factors occurs in the frequency range of roughly 400MHz to 5GHz, and there is great demand for this portion of the radio spectrum.

Importantly, deploying a network that uses higher-frequency capacity bands requires more base stations to cover the same area, and considerably more investment.

Effects of frequency on range

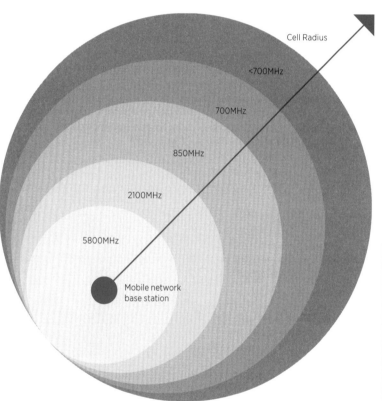

In general, a network that uses higher-frequency spectrum requires more base stations to cover the same area as a network using lower frequencies.

Digital Dividend 1

Background

The Digital Dividend is the spectrum made available for alternative uses following the switchover from analogue to digital terrestrial television, which is more spectrum efficient.

For mobile, the freed-up spectrum has made two potential bands available, 790–862MHz (aka the 800 band) used in ITU-R Region 1 (including Europe, Africa and the Middle East) and 698–806MHz (aka the 700 band) used in ITU-R Region 2 (Americas) and Region 3 (Asia Pacific).

Frequencies below 1GHz are ideal for mobile, offering good geographic coverage, improved in-building coverage, reasonable capacity and availability in large blocks for efficient delivery of mobile broadband.

The Digital Dividend is a key enabler for universal broadband access, bringing socio-economic benefits to people in cities as well as rural and remote areas.

Debate

Which services should Digital Dividend spectrum be licensed for, following the switchover to digital terrestrial television?

What goals should governments try to achieve when relicensing the band?

Broadband networks offer perhaps the greatest opportunity we have ever had to make rapid and solid advances in global social and economic development – across all sectors, including healthcare, education, new job opportunities, transportation, agriculture, trade and government services.

— Houlin Zhao, ITU Secretary-General, January 2015

Industry Position

The Digital Dividend should be allocated to mobile in alignment with regionally harmonised band plans as soon as possible.

The switchover to digital television gives terrestrial broadcasters significantly more capacity for additional channels or high-definition television, even when the Digital Dividend is allocated to mobile.

The economic benefits of licensing the Digital Dividend to mobile are far greater than allocating it to any other service.

Regional harmonisation of the band will permit economies of scale (keeping handset costs low) and mitigate interference along national borders.

Governments should not look to generate excessive fees from the licensing of Digital Dividend spectrum, as this will artificially limit demand, negatively impact network deployment, increase consumer prices and limit the potential economic benefits. Excessive fees can also impede policy goals of delivering broadband access to everyone.

It is reasonable for coverage obligations to be employed to ensure efficient use of this spectrum.

Resources:

GSMA Position Paper: Digital Dividend

GSMA Position Paper: Asia Pacific Digital Dividend/UHF Band Plans

Report: Economic Benefits of the Digital Dividend for Latin America

Report: The Economic Benefits of Early Harmonisation of the Digital Dividend Spectrum and the Cost of Fragmentation in Asia

GSMA Digital Dividend Toolkit

Report: Licensing to Support the Broadband Revolution

Releasing Digital Dividend* Spectrum for Mobile

This map shows individual countries' progress towards the allocation and ultimate licensing of Digital Dividend spectrum for mobile telecommunications.

Digital Dividend spectrum has been licensed to MNOs according to the regionally harmonised band plan.

Digital Dividend regionally harmonised band plan has been announced, not yet licensed to MNOs.

Digital Dividend spectrum has been allocated for mobile – band plan yet to be announced.

Digital Dividend spectrum has not been allocated to mobile.

No information available.

* The Digital Dividend on this map refers to the 800MHz band for Europe, the Middle East
  and Africa, and the 700MHz band for other regions

Source: GSMA Intelligence, November 2015

Digital Dividend 2 Band Plan (EMEA)

Background

In 2015, at the World Radiocommunication Conference in Geneva, agreement was reached to allocate the 694–790 MHz frequency band (aka the 700MHz band) for mobile use in Europe, including Russia, the Middle East and Africa, known as International Telecommunication Union (ITU) Region 1.

This follows a previous agreement to allocate the 703–803 MHz frequency band (also known as the 700MHz band) for mobile services in the Americas and parts of Asia Pacific.

The difference between these two versions of the 700MHz band presents a harmonisation challenge. Specifically, there is a need to agree on a harmonised approach to the band plan in order to drive the economies of scale needed for low-cost consumer devices.

Debate

Because of overlap between the 800MHz band and the Asia Pacific Telecommunity (APT) 700MHz band plan, what should the preferred band plan for the region be?

What is the benefit of a globally harmonised approach to the 700MHz band?

Industry Position

Mobile operators support the proposed 2x30MHz band plan that consists of 703–733MHz (uplink) paired with 758–788MHz (downlink) as the preferred 700MHz band plan for Africa, Middle East and Europe.

This baseline band plan is based on the reuse of the lower duplexer of the APT band plan (i.e., 2 x 30MHz from the APT 2 x 45MHz).

Harmonising the regulatory and technical conditions for the 700MHz band plan in EMEA with the Asia Pacific band plan would maximise economies of scale (keeping handset costs low), mitigate interference along national borders and enable roaming.

Governments should also aim to support the use of the duplex gap for public commercial mobile networks (i.e., supplemental downlink).

However, the mobile industry recognises that some governments may want to consider another option — use of the duplex gap for Public Protection/Disaster Relief (PPDR) mobile broadband applications.

Although governments have options for dedicated PPDR networks outside the 700MHz band, for those that do wish to deploy PPDR within this range, the GSMA recommends that such governmental networks operate outside of the 2x30MHz aligned with the lower duplexer of the harmonised APT band plan.

Resources:
GSMA Public Policy Position on the Preferred Band Plan for Digital Dividend 2 in ITU Region 1
GSMA Welcomes Baltic Governments' Commitment to Harmonising Second Digital Dividend for Mobile Broadband

Deeper Dive

Harmonisation of the Second Digital Dividend in Europe, the Middle East and Africa

The preferred 700MHz band plan for ITU Region 1 aligns with the Asia Pacific Telecommunity (APT) band plan's lower duplexer, offering the potential for near-global harmonisation of the band.

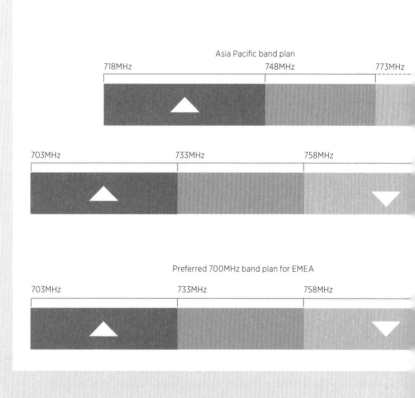

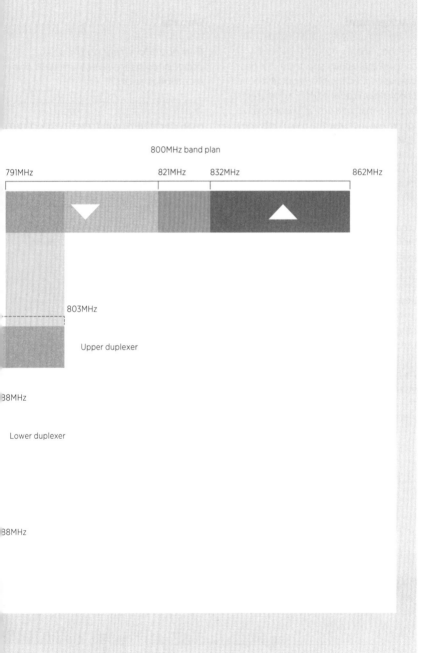

800MHz band plan

791MHz   821MHz   832MHz   862MHz

803MHz

Upper duplexer

38MHz

Lower duplexer

38MHz

Licensed Shared Access

Background

Licensed Shared Access (LSA) is a concept that allows spectrum that has been identified for international mobile telecommunications (IMT) to be used by more than one entity. Theoretically, this would increase the use of the radio spectrum by allowing shared access when and where the primary licensee, a non-mobile incumbent, is not using its designated frequencies.

Licensed shared access complements other authorised ways to access spectrum, including licensed (exclusive) and licence-exempt (unlicensed) use of the spectrum.

Provided that a commercial agreement and an adequate regulatory framework are in place, LSA could allow a portion of assigned spectrum to be used by an LSA user (such as a mobile operator).

As global demand for spectrum intensifies, regulatory strategies such as these are attracting considerable interest and investigation.

Debate

Can operators rely on the LSA concept to share spectrum with the incumbent users?

How can the regulatory/competition issues be addressed with the use of LSA (e.g., to safeguard against one operator getting access to the full LSA spectrum)?

How can LSA be applied effectively, without undermining the urgency of clearing mobile bands for exclusive access?

While we agree that sharing paradigms should be explored as another option for spectrum management, sharing technologies have long promised but remain largely unproven.

— Joan Marsh, Vice President of Federal Regulatory, AT&T

Industry Position

The LSA concept could give mobile network operators access to additional spectrum for mobile broadband, but exclusive access through market-based licensing should remain the main regulatory approach.

LSA does not replace the urgent need to secure additional, exclusive and harmonised spectrum for mobile broadband, and this continues to be the primary objective at the regional and international level.

Authorisation to access additional spectrum using LSA should be granted by national regulatory authorities after public consultation and commercial agreement between the incumbent spectrum user and mobile network operators.

Resources:

The Impact of Licensed Shared Use of Spectrum

GSMA Public Policy Position on Licensed Shared Access (LSA) and Authorised Shared Access (ASA)

Qualcomm: The 1000x Data Challenge

AT&T Public Policy blog: The Power of Licensed Spectrum

Deeper Dive

Spectrum Sharing Models

Licensed use of spectrum, on an exclusive basis, is a time-tested approach for ensuring that spectrum users — including mobile operators — can deliver a high quality of service to consumers without interference. As mobile technologies have proliferated, the demand for access to radio spectrum has intensified, leading to considerable debate and advocacy for new approaches to spectrum management.

Licence-exempt spectrum:

Frequency bands that can be used by multiple systems and services if they meet predefined 'politeness protocols' and technical standards. Wi-Fi is a technology that uses licence-exempt spectrum.

Shared licensed spectrum:

Any licensed spectrum that is shared among licensed users. This sharing may be agreed on a commercial basis between licensed entities or as a condition of the licensing process.

TV white space:

Television spectrum in the UHF band that, due to predictable geographical or temporal gaps in broadcasting, offers the potential for licence-exempt devices to use the spectrum for broadband services. These services are dependent on dynamic spectrum management technologies and techniques.

Licensed shared access (or authorised shared access):

A proposed sharing scheme that allows licensed use of underutilised spectrum that is already licensed by another service. Licensed shared access (LSA) is proposed as a way to ensure a high quality of service is delivered, as opposed to best-endeavour services that are delivered through licence-exempt spectrum.

While these innovations may find a viable niche in the future, the GSMA's position is that pursuit of these options today risks deflecting attention from the release of sufficient, exclusively licensed spectrum for mobile broadband.

Limiting Interference

Background

Radio transmissions always have the potential to interfere with radio systems operating in adjacent frequency bands, due to transmitter imperfections or imperfect receiver filtering.

New technologies are better at mitigating interference than in the past, although they can be more costly due to equipment complexity and energy consumption.

The solution is to define radio transmitter and receiver parameters to ensure compatibility between radio systems operating in the same or adjacent frequency bands. This approach cannot, however, be applied to technologies that lack standards.

The traditional way to manage interference has been to establish guard bands that are left vacant. However, these guard bands reduce the overall efficiency of spectrum use. Other interference-mitigation techniques should be employed as much as possible to minimise the loss of usable spectrum.

Debate

Are guard bands the only way to prevent interference between mobile bands and systems using adjacent bands?

Should potential interference be solved ex-ante by the national regulatory authority before allocating new spectrum to mobile operators, or should this be left to the operators?

The more countries that support a band, the greater the possibility for global harmonisation, offering substantial economies of scale, reducing interference along country borders and delivering cost benefits for consumers.

— GSMA

Industry Position

Interference can be managed with proper planning and mitigation techniques.

For mobile telecommunications, regional harmonisation of allocated mobile bands is the best way to avoid interference along national borders.

Issues of cross-border interference are usually addressed through bilateral or multilateral agreements among neighbouring countries.

To minimise guard band size and the cost of interference mitigation, radio system standards defining transmitter and receiver RF performance are necessary.

Broadcasters are rightly concerned that mobile services introduced in the UHF band do not interfere with television reception, and mobile operators are equally concerned that this does not happen. A television receiver standard would improve the situation.

Resources:
Technical paper: Managing Radio Interference
GSMA briefing paper on WRC Agenda Item 1.17 — broadcast interference
Fact sheet: Potential for Interference to Electronics

Real-World Experience of 800MHz LTE Coexistence

Because Digital Dividend spectrum is, by definition, adjacent to frequency bands that continue to be used for television broadcasting, regulators and industry have worked hard to ensure that mobile service using the 800MHz Digital Dividend band does not interfere with television broadcasting. Nevertheless, concerns continue to be aired in most markets until the actual roll out of the mobile service. Now that mobile network operators in several countries have begun to deploy LTE networks using Digital Dividend spectrum, these concerns can be largely put to rest.

In Germany, as of October 2012, more than 4,600 800MHz base station sites had been deployed, in urban, suburban and rural areas. Reported incidents of interference were very low. Six cases of interference with digital terrestrial television were reported, and this includes the most critical case, involving the lower block of LTE spectrum and TV channel 60, which O2 rolled out in Nuremburg in July 2012. In addition, 22 cases involved wireless microphones (which had already been asked to migrate to other frequencies by the regulator), and six involved other radio services and applications.

In Sweden, hundreds of 800MHz base station sites have been deployed, with the first-line response for reported interference managed jointly by the mobile operators. During the first quarter of 2012, approximately 40 cases of interference with the television bands were reported, of which 30 were quickly resolved by supplying the viewers with a television receiver filter.

Globally, up to now, there have been fewer cases of interference with digital terrestrial television by mobile services in the 800MHz band than was forecast. However, the incidence rate may vary depending on the proportion of the population that uses the digital television platform and the digital television network topology. Radio frequency (RF) amplifiers are a more significant factor than anticipated, but RF filters can solve the majority of interference cases. So far, there has been no interference to cable networks.

Source: Vodafone

at800 in the United Kingdom

In 2012, mobile operator licensees in the UK set up a joint venture called at800 to act as the mechanism for resolving television interference issues when LTE services were launched in the 800MHz band.

The four mobile operators are shareholders, and each had to contribute £30m per 5MHz lot acquired. at800 was then responsible for collecting information about each operator's LTE800 roll out plans and arranging a leafleting campaign in the affected areas, giving details of how householders could report interference issues. at800 manages the call centre, posts filters to consumers and sends engineers to fix any remaining problems. Any funds remaining after the completion of the programme will be divided among the shareholders. In practice, it has become apparent that the scale of interference was greatly overestimated.

As of 6 May 2015, at800 had handled around 227,500 calls from viewers and responded to just over 5,500 people on social media. For viewers experiencing disruption that is not related to LTE at 800MHz, at800 directs viewers to organisations that may be able to help.

Spectrum Auctions

Background

Spectrum management for mobile telecommunications is increasingly complex as governments release new spectrum in existing mobile bands, manage the renewal of licences coming to the end of their initial term, and release spectrum in new bands for mobile broadband services.

Effective and efficient management of these processes is central to the continued investment in, and development of, mobile services.

Auctions are an efficient way to allocate spectrum when there is competition for scarce spectrum resources and demand is expected to exceed supply.

There are a number of alternative auction designs, each with its strengths and limitations. While multi-round auctions are often preferred, the best choice is dependent on the market circumstances and the objectives of the government and regulators.

When assigning spectrum via an auction, governments typically have a number of goals, which may include achieving:

- The maximum long-term value to the economy and society from the use of the spectrum.

- Efficient technical implementation of services.

- Sufficient investment to roll out networks and new services.

- Revenue generation for the government.

- Adequate market competition.

- A fair and transparent allocation process.

Debate

How is the value of spectrum best determined?

What are the main considerations for auction design, to achieve the government's desired outcomes?

Should governments design auctions to maximise revenue in the short term, or to ensure an economically efficient means of allocating a scarce resource?

The countries that get their spectrum policy right will achieve widespread access to affordable and innovative mobile broadband services. Strong communications infrastructure, in turn, brings significant wider economic benefits including in boosting productivity and living standards.

— Competition Economists Group, 2012

Industry Position

Efficient allocation of spectrum is necessary to realise the full economic and societal value of mobile.

There is no 'one size fits all' design for spectrum auctions. Each auction needs to be designed to meet the market circumstances and to achieve the specific objectives set by government.

As with most auction design elements, the appropriateness of simultaneous auctions (multiple bands being auctioned together) versus sequential auctions (bands being auctioned one after the other) is dependent on specific market conditions. The effectiveness of either approach will be dependent on a clear spectrum road map with well-defined rights and conditions understood in advance.

Regulators should work with stakeholders to ensure the auction design is fair, transparent and appropriate for the specific market circumstances. Auctions are not the only option available to governments to manage spectrum allocation and should only be used in appropriate circumstances.

Auctions should be designed to maximise the long-term economic and social benefits from use of the spectrum. They should not be designed to maximise short-term revenue for governments.

Resources:
Report: Licensing to Support the Broadband Revolution
GSMA Position Paper: Spectrum Auctions
GSMA Position Paper: Spectrum Licensing

Case Study

Reserve Pricing for Spectrum Auctions

Reserve prices play an important role in spectrum auction design. They discourage non-serious bidders and can also ensure that a minimum price is paid for spectrum licences when competition for the spectrum is weak. When competition for access to mobile spectrum is anticipated to be strong, however, it does not follow that high reserve prices should be set. In fact, it risks alienating potential bidders and could lead to auction failure, leaving valuable spectrum unsold and unused.

Rather than focusing on revenue maximisation, governments would be wiser to focus on the positive social and economic outcomes generated by widespread mobile service, while assuring an appropriate level of industry competition. Lower, realistic reserve prices for spectrum auctions allow the market to determine the value of the spectrum being released. Following are two auctions where reserve prices played a critical role:

India: Hooked on high reserve prices

In March 2013, Indian telecommunications regulator TRAI conducted an auction of 1800MHz spectrum in four of its national 'circles' as well as 900MHz spectrum in three circles and 850MHz spectrum as a pan-Indian licence. Industry response to the offering was poor, as the reserve prices were deemed to be very high given the nature of the market, with its low consumer tariffs. Reserve prices for the 900MHz lots were set at twice the reserve prices of the 1800MHz spectrum in the same circles, for example. In the end, the auction attracted only one bidder, MTS, which secured 850MHz spectrum for just eight of the 22 service areas.

Australia: The first Digital Dividend spectrum to be left unsold

In May 2013, Australia's auction of Digital Dividend spectrum concluded, leaving one-third of the 700MHz band unsold. The auction, which also included lots of 2.6GHz spectrum, generated AUD$1 billion ($780 million) less than the government had predicted. It was reportedly the first occurrence of any Digital Dividend spectrum being left unsold. The Australian government has since come under fire for setting the reserve price unrealistically high at $1.43/MHz/population. Of the country's three incumbent mobile operators, Telstra and Optus bought less of the 700MHz spectrum than they were allowed to under the auction rules, and Vodafone Hutchison Australia chose not to bid at all.

In the words of Brett Tarnutzer, Head of Spectrum, GSMA, "Acquiring spectrum is only the first step before making the necessary investment in network deployment to deliver mobile services to consumers. Unreasonably high reserve prices lead to spectrum remaining unsold, delays in the delivery of mobile services and, ultimately, an increase in consumer tariffs."

Spectrum Caps

Background

Spectrum caps are limits to how much spectrum can be licensed by any mobile operator. They are used by governments and regulators to manage the allocation of spectrum during auctions. The intention is to ensure effective competition and to prevent existing operators from using their economic strength to secure large spectrum assets, which could give them a competitive advantage in the future.

Spectrum caps are increasingly used by regulators in auction rules to encourage spectrum reallocation and to balance operator portfolios.

New entrants and players with fewer spectrum assets typically support caps on new spectrum allocations, while incumbents argue that the approach negatively impacts the quality of service they can deliver to consumers.

Debate

Does the use of caps in spectrum allocation result in the best social and economic outcomes?

Are spectrum caps an appropriate way to address market dominance?

Industry Position

In markets where competition is ineffective, the use of spectrum caps may be appropriate, but care must be taken to avoid unintended consequences and poor outcomes for consumers.

Operators should not be penalised for using their spectrum assets successfully or constrained in delivering new services. Operators with the largest market share are usually the ones that need more spectrum to meet customer demand.

Spectrum caps, when applied without discrimination among the operators, distribute spectrum among market players and, potentially, new entrants. If imposed, they should allow all operators to deploy networks in a technically and economically efficient manner.

Auction and licensing rules must give operators the opportunity to secure a portfolio of spectrum to deliver economically viable broadband services.

Using spectrum caps specifically to attract new market entrants can lead to spectrum fragmentation and market inefficiencies which, ultimately, will negatively affect consumers and businesses using mobile services. Licence conditions for network deployment and spectrum use may lead to more effective outcomes for consumers.

Before applying spectrum caps, regulators should conduct a rigorous market analysis to ensure there are, in fact, other operators in the market whose access to spectrum would deliver greater societal benefits.

Market dominance should not be addressed through spectrum caps, but through antitrust measures.

Resources:

Report: Licensing to Support the Broadband Revolution

Report: Mobile Broadband, Competition and Spectrum Caps

Article: Forbes.com, 'Sending the Wrong Signals to the Wireless Marketplace'

Assessing the Impact of Spectrum Caps in Chile

In September 2009, Chilean regulator Subtel licensed 90MHz of the AWS (1.7–2.1GHz) spectrum band, divided into three blocks, for national mobile service. In doing so, Chile became the first Latin American country to license this band.

The Chilean Supreme Court authorised Subtel to impose a spectrum cap of 60MHz, effectively excluding the three incumbent mobile network operators — Movistar (Telefónica), Entel and Claro (América Móvil) — all of which were at or near the 60MHz threshold with their existing spectrum portfolios.

Cable television company VTR won block A of the AWS spectrum with an offer of US$3.02 million, and Nextel (rebranded as WOM in 2015) won blocks B and C, paying US$14.7 million. Both operators were required to deploy services within one year.

"The entry of two new companies will increase competition in the mobile phone and internet business, which is good news for 15 million Chileans," transport and telecommunications minister René Cortázar told reporters at the time.

With the benefit of hindsight, was the spectrum cap an effective strategy to increase competition and benefit citizens? Not entirely. Despite the requirement of a swift roll out of services, the new entrants were unable to launch their 3G mobile service until May 2012, one-and-a-half years after the October 2010 deadline.

Nor has the competitive landscape been dramatically altered. By August 2013, VTR and Nextel together had only achieved a 1.3 per cent market share. Later in 2013, VTR stopped using its own network and became a Mobile Virtual Network Operator through an agreement with Movistar. By September 2014, the government had sent a draft bill to Congress for the development of a secondary market to enable spectrum trading, as some companies were not making use of all of their spectrum.

Assessing the Impact of Spectrum Caps in Chile

Company	Subscribers		Market Share*		Spectrum Holidays	
	Q3 2009	Q2 2013	Q3 2009	Q2 2013	Before	After
Entel	6,126,037	10,141,135	36.64%	37.36%	60MHz (35.0%)	60MHz (23%)
Claro	3,302,000	6,275,000	19.75%	23.12%	55MHz (32.5%)	55MHz (21%)
Movistar	7,255,400	10,377,100	43.39%	38.23%	55MHz (32.5%)	55MHz (21%)
Nextel (rebranded as WOM)	38,000	208,100	0.23%	0.77%	–	60MHz (23%)
VTR**	–	140,100	–	0.52%	–	30MHz (12%)

*The 2.6GHz band was licensed in June 2012 (40MHz for Entel, 40MHz for Claro and 40MHz for Movistar).

**VTR switched to operating as a Mobile Virtual Network Operator in 2014 through an agreement with Movistar.

Source: GSMA Intelligence

Spectrum Harmonisation

Background

Spectrum harmonisation refers to the uniform allocation of radio frequency bands, under common technical and regulatory regimes, across entire regions. A country's adherence to internationally identified spectrum bands offers many advantages:

- Lower costs for consumers, as device manufacturers can mass-produce devices that function in multiple countries on a single band.

- Availability of a wider portfolio of devices, driven by a larger, international market.

- Roaming, or the ability to use one's mobile device abroad.

- Fewer issues of cross-border interference.

There are a limited number of bands that can be supported in a mobile device. Each new band supported increases the device cost, reduces the receiver's sensitivity and drains the battery.

Harmonised bands have enabled huge economies of scale, leading to unprecedented use of mobile telecommunications worldwide. Spectrum bands for international mobile telecommunications (IMT) are defined through a rigorous multilateral process that considers their technical and practical merits.

In 2015, at the World Radiocommunication Conference (WRC) in Geneva, agreement was reached on the creation of three global spectrum bands for mobile —700MHz, 1427–1518MHz and 3.4–3.6GHz. The outcome provides the industry with an important mix of internationally harmonised coverage and capacity spectrum to meet growing demand for mobile services. Spectrum harmonisation through the WRC process is also key to enabling lower cost mobile devices through economies of scale.

The global harmonisation of the 694–790MHz frequency band that has been decided by WRC-15 paves the way for manufacturers and mobile operators to offer mobile broadband at an affordable price in currently underserved areas.

— François Rancy, Director, ITU Radiocommunication Bureau

Debate

*How harmonised does a band
need to be to realise the benefits
of harmonisation?*

---

*Can a national market be so large
that the benefits of spectrum
harmonisation are inconsequential?*

---

*In the future, will cognitive
technologies enable devices to tune
dynamically to any band removing
the need for countries to harmonise?*

Industry Position

**Governments that align national use
of the spectrum with internationally
harmonised band plans will achieve the
greatest benefits for consumers and
avoid interference along their borders.**

At a minimum, harmonisation of mobile
bands at the regional level is crucial. Even
small variations on standard band plans
can result in device manufacturers having
to build market-specific devices, with
costly consequences for consumers.

All markets should harmonise regionally
where possible, as this benefits the entire
global mobile ecosystem. There is no
advantage to going it alone.

Cognitive radio technologies will not reduce
the need for harmonised mobile spectrum
anytime soon. Adhering to internationally
recognised band plans is the only way to
achieve large economies of scale.

Resources:

Report: The Economic Benefits of Early Harmonisation of the Digital Dividend Spectrum
and the Cost of Fragmentation in Asia

Report: The Benefits of Releasing Spectrum for Mobile Broadband in Sub-Saharan Africa

Report: Economic Benefits of the Digital Dividend for Latin America

Spectrum Licence Renewal

Background

Many of the original 2G spectrum licences are coming up for renewal in the next few years. National regulatory authorities must determine how mobile operators' spectrum rights will be affected as licences approach the end of their initial term.

The prospect of licence expiry creates significant uncertainty for mobile operators. A transparent, predictable and coherent approach to renewal is therefore important, enabling operators to make rational, long-term investment decisions.

There is no standard approach to relicensing spectrum. Each market needs to be considered independently, with industry stakeholders involved at all stages of the decision process. Failure to effectively manage the process can delay investment in new services and affect mobile services for, potentially, millions of consumers.

Debate

Which approach to spectrum licence renewal will have the most beneficial outcome for consumers and society?

Should spectrum licence holders presume they will have the option to renew when the licence reaches the end of its term, unless otherwise specified in the licence?

Should governments feel free to reshuffle spectrum allocations, change bandwidths or alter licence conditions on renewal?

Industry Position

It is essential that governments and regulators implement a clear and timely process for the renewal of spectrum licences.

Maintaining mobile service for consumers is critical. To ensure this, the approach for licence renewal should be agreed at least three to four years before licence expiry.

Governments and regulators should work on the presumption of licence renewal for the existing licence holder. Exceptions should only apply if there has been a serious breach of licence conditions in advance of renewal.

Should a government choose to reappraise the market structure at the time of renewal, the priorities should be to maintain service for consumers and ensure network investments are not stranded. Governments should not discriminate in favour of, or against, new market entrants, but establish a level playing field.

New licences should be granted for 15 to 20 years, at least, to give investors adequate time to realise a reasonable return on their investment.

Renewed mobile licences should be technology and service neutral.

Resources:
Position Paper: Renewal of Spectrum Usage Rights
Report: Licensing to Support the Mobile Broadband Revolution

Spectrum Licensing

Background

Spectrum licensing is a powerful lever that national regulatory authorities can use to influence the competitive structure and behaviour of the mobile telecoms sector.

The amount of spectrum made available and the terms on which it is licensed fundamentally drive the cost, range and availability of mobile services.

Mobile is a capital-intensive industry requiring significant investment in infrastructure. Governments' spectrum licensing policy — when supported by a stable, predictable and transparent regulatory regime — can dramatically raise the attractiveness of markets to investors.

Spectrum management for mobile telecommunications is complex, as governments release new spectrum in existing mobile bands; manage the renewal of licences coming to the end of their initial term; and release spectrum in new bands for mobile broadband services.

Debate

What is the most effective way to license spectrum?

What conditions should be tied to spectrum access rights?

Are licensing rules the best way to ensure a healthy, well-functioning mobile sector, or should the development of the industry be shaped predominantly by market forces?

Industry Position

Spectrum rights should be assigned to the services and operators that can generate the greatest benefit to society from the use of that spectrum.

Regulatory authorities should foster a transparent and stable licensing framework that prioritises exclusive access rights, promotes a high quality of service and encourages investment.

Licensing authorities should publish a road map of the planned release of additional spectrum bands to maximise the benefits of spectrum use. The road map should take a five- to ten-year view and include a comprehensive and reasonably detailed inventory of current use.

Restrictive licence conditions limit operators' ability to use their spectrum resources fully, and risk delaying investment in new services.

In particular, service and technology restrictions in existing licences should be removed.

To the maximum practical extent, spectrum should be identified, allocated and licensed in alignment with internationally harmonised mobile spectrum bands to enable international economies of scale, reduce cross-border interference and facilitate international services.

For new spectrum allocations, market-based approaches to licensing, such as auctions, are the most efficient way to assign spectrum to the bidders that value the spectrum the most.

Licence fees should be used to help recover the administrative costs of freeing up spectrum for new, higher-value uses, and licensing and managing the spectrum for long-term social and economic benefit. They should not be used to maximise government revenue.

Resources
Report: Licensing to Support the Broadband Revolution
GSMA Position Paper: Spectrum Licensing

Spectrum Trading

Background

Spectrum trading is a mechanism by which mobile network operators can transfer spectrum usage rights on a voluntary commercial basis.

Trading spectrum usage rights is a relatively recent development. In Europe, most countries that allow the practice have done so since 2002 or later, and each country has established different rules governing the practice.

Trading rules can facilitate the partial transfer of a usage right, which could permit a licensee to use a specified frequency band at a particular location or for a certain duration. This may result in more intensive use of the limited spectrum.

Debate

Should spectrum-trading arrangements between mobile network operators be allowed?

What role should regulators play in overseeing such arrangements?

What regulatory procedures are required to ensure transparency and notification of voluntary spectrum trading?

Industry Position

Countries should have a regulatory framework that allows operators to engage in voluntary spectrum trading.

Spectrum trading creates increased flexibility in business planning and ensures that spectrum does not lie fallow, but instead is used to deliver valuable services to citizens.

Spectrum trading restrictions should only be applied when competitive or other compelling concerns are present.

Spectrum trading agreements are governed by commercial law and subject to the rules applicable to such agreements. They may also be subject to assessment under competition law.

It makes sense for governments to be notified of spectrum trading agreements and to grant approval. Notification requirements preserve transparency, making it clear which entities hold spectrum usage rights and ensuring that trading arrangements are not anti-competitive.

Governments should implement appropriate and effective procedures for handling notification requests of spectrum trading agreements.

Resources
Position Paper: Spectrum Trading
GSMA Europe consultation response: Secondary trading of rights to use spectrum
CEPT/CEE Report: Description of Practices Relative to Trading of Spectrum Rights of Use

Spectrum Trading in Guatemala

Guatemala is one of the few countries that permits spectrum trading and where the practice is widespread. In 1996, the Guatemalan government chose to allow spectrum trading in specific liberalised frequency bands. This did not apply to bands that were allocated nationally for government or private amateur radio use, to protect spectrum for vital public services and individuals.

However, the bands allocated to commercial applications such as broadcasting and mobile services were liberalised, allowing licences lasting for 15 years to be leased, sold, subdivided or aggregated at the owner's discretion — and renewed for a longer period on request.

This kind of licence, known as a Título de Usufructo de Frecuencia (TUF), permits use over a specific frequency range in a certain geographical area at certain times, and is subject to power restrictions to prevent interference, especially close to national borders.

As such, the role of the regulator is restricted to adjudicating over interference disputes where mediation has failed, as well as managing non-liberalised government spectrum.

The TUF allocation process:

Interested parties submit formal requests, to which the government must publicly respond within three days.

Third parties have five days to oppose the request.

The only reasons requests may be denied are for violation of an international treaty (surrounding use of the frequency band) or if the existing right to flat frequency range is already held by another.

Assuming the requests meet these criteria, an auction must be announced within 15 days and must take place within 20 days after that.

Technology Neutrality and Change of Use

Background

Technology neutrality is a policy approach that allows the use of any non-interfering technology in any frequency band.

In practice, this means that governments allocate and license spectrum for particular services (e.g., broadcasting, mobile, satellite), but do not specify the underlying technology used (e.g., 3G, LTE or WiMAX).

Many of the original mobile licences were issued for a specific technology, such as GSM or CDMA, which restricts the ability of the licence holder to 'refarm' the band using an alternative, more efficient technology.

Refarming refers to the repurposing of assigned frequency bands, such as those used for 2G mobile services (using GSM technology) for newer technologies, including third-generation (UMTS technology) and fourth-generation (LTE technology) mobile services.

Spectrum allocations for International Mobile Telecommunications (IMT) are technology-neutral. IMT technologies including GPRS, EDGE, UMTS, HSPA, LTE and WiMAX are standardised for technical coexistence.

Debate

Should governments set the technical parameters for a band's use or should the market decide?

Should licence conditions restrict operators' ability to deploy more efficient technologies and adapt to market changes?

How is spectrum coexistence best managed to prevent interference between services and operators using different technologies?

We know that the choice of the wrong standard can lock our economies into long periods of economic underperformance, while market-led solutions have consistently provided a much better environment for technology selection.

— European Commissioner Viviane Reding, 4 December 2006

Industry Position

We support a licensing approach that allows any compatible, non-interfering technology to be used in mobile frequency bands.

Adopting harmonised, regional band plans for mobile ensures that interference between services can be managed. Governments should allow operators to deploy any mobile technology that can technically coexist within the international band plan.

Technology neutrality encourages innovation and promotes competition, allowing markets to determine which technologies succeed, to the benefit of consumers and society.

Governments should amend technology-specific licences to allow new technologies to be deployed, enabling operators to serve more subscribers and provide each subscriber with better, more innovative services per unit of bandwidth.

Enabling spectrum licence holders to change the underlying technology of their service, known as refarming, generates positive economic and social outcomes and should be allowed.

Resources:
Position Paper: Change of Use of Spectrum
Report: Licensing to Support the Broadband Revolution

Deeper Dive

The 1800MHz band: a global refarming success story for LTE

The lack of truly global LTE frequency bands made it difficult to establish a wide range of low-cost devices for the first phase of 4G services. It also prevented widespread international roaming.

Because mobile devices can only support a limited number of frequency bands, a lack of harmonised bands means devices can only operate and be sold in a limited number of markets. This problem was highlighted when several early 4G-enabled Apple devices could not operate on some 4G networks around the world, as they did not support the right frequency bands.

A critical part of the solution has been the 1800MHz band, which has traditionally been used for 2G GSM services. The band has historically been one of the key enablers of low-cost devices and international roaming, as it is one of the only bands to be harmonised worldwide.

In countries where regulators support technology-neutral spectrum licences, operators have been able to refarm the 1800MHz band for LTE services. The 1800MHz band is now the most widely deployed LTE band globally, as well as the most widely supported in mobile devices. According to the Global Mobile Suppliers Association (GSA), the 1800MHz band has the largest device ecosystem of any LTE band with over 1,543 compatible user devices available as of October 2015.

Mapping 4G-LTE deployments by frequency bands

420 operators worldwide now have live LTE networks, covering 132 countries. As many operators use multiple spectrum bands in their LTE networks, this equates to more than 600 individual deployments.

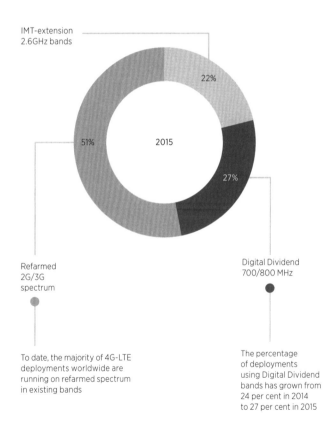

Percentage of spectrum scenarios used
in global 4G-LTE deployments

IMT-extension
2.6GHz bands

22%

2015

51%

27%

Reafarmed
2G/3G
spectrum

Digital Dividend
700/800 MHz

To date, the majority of 4G-LTE deployments worldwide are running on reafarmed spectrum in existing bands

The percentage of deployments using Digital Dividend bands has grown from 24 per cent in 2014 to 27 per cent in 2015

4G-LTE deployments by frequency bands, October 2015

Source: GSMA Intelligence

163

TV White Space

Background

The expression 'white space' is used to define the parts of the spectrum that are not used at a given time and geographical location.

Typically, TV white space consists of unused spectrum in the television broadcasting bands (e.g., 470–790MHz in Europe and 470–698MHz in the United States).

There is unused spectrum in these bands mainly because of the necessary geographical separation between television stations of the same channel, as well as parts of the spectrum dedicated to regional television stations that remain unused in certain areas.

Some internet players are advocating globally for use of TV white space for licence-exempt services such as Wi-Fi. It is worth noting that commercially desirable geographic areas, such as major urban and suburban areas with high population and business densities, typically have little, if any, TV white space at all.

Debate

What kinds of applications can take advantage of TV white space?

In reality, how much TV white space is available?

What licensing regime is most appropriate to get the maximum benefit from spectrum resources for mobile broadband?

The over-eager pursuit of unlicensed sharing models cannot turn a blind eye on the model proven to deliver investment, innovation, and jobs – exclusive licensing. Industry and government alike must continue with the hard work of clearing and licensing under-utilised government spectrum where feasible.

— Joan Marsh, Vice President of Federal Regulatory, AT&T

Industry Position

Use of TV white space must not jeopardise the future of the UHF band, especially in the case of reallocation for exclusive mobile use.

The use of TV white space must not distort the market through inappropriate regulation. Eliminating the cost of acquiring licensed spectrum to provide cellular-type mobile services could create an unfair advantage.

The TV white space approach is made possible by a spectrum-use database including geo-location data, which cannot offer a predictable quality of service or spectrum availability. For TV white space, there is no *a priori* determination of the spectrum to be eventually accessed.

Interference management remains a top priority. The use of TV white space, on a secondary unlicensed basis, requires careful avoidance of interference with primary users such as existing TV broadcasters, as well as services in adjacent bands.

It is important to consider how to use the Digital Dividend spectrum most effectively to benefit citizens and businesses, and discussions about TV white space should not derail this process.

Resources:
GSMA Public Policy Position on TV White Space
GSMA Europe response to Radio Spectrum Policy Group 2010 Work Programme
AT&T Public Policy Blog: The Power of Licensed Spectrum

Consumer Protection

For a vast number of people, mobile phones serve as a personal portal to the friends, family, services and resources they rely on every day. It is essential for the mobile industry, therefore, to deliver safe and secure technologies — complemented by safe and secure mobile apps — that inspire trust and confidence. At the same time, consumers need to be aware of their role in avoiding risks.

Mobile technologies are not immune to the issues faced offline and by other forms of information and communication technology. For example, criminal activities such as online exploitation of children, spamming and device or identity theft existed before the proliferation of mobile technologies.

The mobile industry takes consumer protection seriously. The GSMA and its members work with governments, multilateral organisations and non-governmental organisations to address mobile-related threats to citizens by:

- Commissioning research that offers real-world insight and evidence.

- Building and participating in cross-sector coalitions.

- Defining and promoting global best practice.

- Leading technical initiatives.

The following pages provide a small indication of the work undertaken by the mobile industry to ensure consumers are appropriately protected and informed as they enjoy the full range of benefits that mobile technology makes possible.

Children and Mobile Technology

Background

Young children and teenagers are enthusiastic users of mobile technology. The 2013 report Children's Use of Mobile Phones — An International Comparison reveals that 81 per cent of children aged 8–18 in the countries surveyed use a mobile phone, and 55 per cent of those children use their mobile phone to access the internet. Young people's knowledge of mobile applications and platforms often surpasses that of parents, guardians and teachers, and the international comparison report confirmed that children use social networking services more than their parents.

For growing numbers of young people, mobile technology is an increasingly important tool for communicating, accessing information and entertainment, learning, playing and being creative. As mobile technology becomes increasingly embedded into everyday life, mobile phone operators can play an important role in protecting and promoting children's rights.

Mobiles can be key enablers to access:

- Skills for employment.

- Enhanced formal and informal education and learning.

- Information and services to aid in health and well-being.

- Improved social and civic engagement.

- Opportunities to play and to be creative.

Mobile devices increasingly play a role in formal education and informal learning. In developing and rural areas, as well as places where certain people — girls in particular — are excluded from formal education, mobile connectivity offers new opportunities to learn.

Like any tool, mobile devices can be used in ways that cause harm, so children require guidance and a safe, secure environment to benefit from mobile technologies.

The mobile industry has taken active steps in the area of child online protection. The GSMA has played a leading role in self-regulatory initiatives dealing with issues such as parental controls, education and awareness.

Debate

What potential harms are children exposed to in the online environment?

To what extent can technology protect young people from online threats, and what role does consumer awareness and education play?

We are grateful for the leadership shown by the GSMA Mobile Alliance members in tackling online child sexual abuse material. Their coordinated action helps set the standard and illustrates how proactive steps taken by industry can help protect children's rights in today's digital society.

— Eija Hietavuo, CSR Manager, UNICEF

Is industry doing enough to protect children when they are online, and what is the role of parents and teachers?

Should governments require mobile operators, through regulation, to take steps to protect children from online risks?

Are concerns about online risks preventing mobile learning and education opportunities from being fully realised?

Industry Position

Mobile devices and services enhance the lives of young people. This perspective needs to be embraced, encouraged and better understood by all stakeholders to ensure young people get the maximum benefits from mobile technology.

Addressing child online protection is best approached through multi-stakeholder efforts. The GSMA takes part in international initiatives related to child online protection, including contributing to the ITU's Child Online Protection programme, and actively engages with governments and regulators looking to address this issue.

Working closely with UNICEF, the GSMA and its mobile operator members, as well as a range of other organisations including the International Centre for Missing and Exploited Children (ICMEC), INHOPE and INTERPOL, hold national and regional multi-stakeholder workshops on the issue. These workshops bring together policy makers, NGOs, law enforcement and industry, to facilitate the development of collaborative approaches to safe and responsible use of the internet.

Through its mYouth programme, the GSMA also leads several initiatives to promote the safe use of mobile services for young people, provides useful research on child online safety, and gathers evidence about how young people use their mobile devices in different parts of the world.

Young people are critical to the evolution of the mobile sector as they represent the first generation to have grown up in a connected, always-on world. They are future consumers and innovators who will deliver the next wave of innovation in mobile.

Resources:

UNICEF: Guidelines for Industry on Child Online Protection
European Framework for Safer Mobile Use
ICT Coalition
GSMA: mYouth
GSMA Report: Children's Use of Mobile Phones, An International Comparison 2013
GSMA Report: Children's Use of Mobile Phones, An International Comparison 2012

Children's Use of Mobile Phones in Algeria, Egypt, Iraq and Saudi Arabia

Since 2008, the GSMA has been collaborating with NTT DOCOMO's Mobile Society Research Institute on a multiyear project to better understand how children aged eight to 18 use mobile phones around the world.

The research is comparative, typically covering four or five different countries. Some standard questions, posed to children and their parents since the beginning of the programme, enable broad year-on-year comparisons on areas such as age of first mobile ownership and the reasons for getting a phone, as well as parents' concerns about their children's use of mobile. New questions are added to account for the evolution of children's mobile lives – more recent research, for example, has asked children about accessing social media services from mobiles and how they manage their privacy settings.

Countries taking part in the research are able to develop a targeted understanding of the real mobile habits of younger users and can therefore develop strategies for promoting safe and responsible use of mobile from a firmer foundation.

55% of all child mobile phone users access the mobile internet This increases to **93%** when looking exclusively at child smartphone users

Over **60%** of parents have concerns about children's mobile phone use, with viewing inappropriate sites the highest percentage at **85%**

57% of parents who have access to parental control solutions used them; content filters are the most popular control method at **56%**

87% of children surveyed say that having a mobile phone increases their confidence; this is particularly the case in Saudi Arabia where the figure rises to **98%**

75% of parents believe that an adult in the family should educate their children about mobile phone use; this is a consistent preference across all countries

91% of function use is camera features, **88%** music players and **78%** movie players

63% of all children who use the internet through their mobile phone access it between one and five times a day, with **21%** accessing it more than six times a day and only **16%** accessing it less than once a day

40% of children on social networking sites have public profiles, though girls are more likely than boys to have private profiles

Of those children who access the internet via their smartphones... **85%** of them download or use apps

55% More than half of all child mobile phone users surveyed make use of location based services

Children use social networking services more than their parents across all four continents

72% of children who use social networking service communicate with 'new friends' online

73% of parents surveyed expressed concern about their children's privacy when using mobile phones, with equal concern expressed for girls and boys

Source: GSMA and NTT DOCOMO

About the ICT Coalition

The ICT Coalition for the Safer Use of Connected Devices and Online Services by Children and Young People in the EU (www.ictcoalition.eu) is made up of 23 companies from across the information and communication technology (ICT) sector. Members of the ICT Coalition pledge to encourage the safe and responsible use of online services and internet devices among children and young people and to empower parents and carers to engage with and help protect their children in the digital world.

The principles are suitably high level, enabling their application to evolve as technology and consumer propositions evolve, and to facilitate their adoption by a variety of companies and services. The ICT Coalition's members include leading internet and online service providers such as Google and Facebook, device manufacturers, and mobile operators including Deutsche Telekom, KPN, Orange, Portugal Telecom, TDC, Telecom Italia, Telefónica, Telenor, TeliaSonera and Vodafone.

Members of the ICT Coalition are required to specify how their organisation will deliver on six principles related to online content, parental controls, dealing with abuse and misuse, child abuse and illegal contact, privacy and control, and education and awareness.

Electromagnetic Fields and Device Safety

Background

According to the World Health Organization (WHO), there are no established health risks from the radio signals of mobile devices that comply with international safety recommendations.

However, research has shown a possible increased risk of brain tumours among long-term users of mobile phones. As a result, in May 2011, radio signals were classed as a possible human carcinogen by the International Agency for Research on Cancer. Health authorities have advised that given scientific uncertainty and the lack of support from cancer trend data this classification should be understood as meaning that more research is needed. They have also reminded mobile phone users that they can take practical measures to reduce exposure, such as using a handsfree kit or text messaging.

Mobile phone compliance is based on an assessment of the specific absorption rate (SAR), which is the amount of radiofrequency (RF) energy absorbed by the body.

Mobile phones use adaptive power control to transmit at the minimum power required for call quality. When coverage is good, the RF output level may be similar to that of a home cordless phone.

Some parents are concerned about whether mobile phone use or the proximity of base stations to schools, daycare centres or homes could pose a risk to children. National authorities in some countries have recommended precautionary restrictions on phone use by younger children, while others, such as the US Food and Drug Administration (FDA), have concluded that current scientific evidence does not justify measures beyond international safety guidelines.

A comprehensive health risk assessment of radio signals, including those of mobile phones, is being conducted by the WHO. The conclusions are expected in 2016.

Debate

*Is there a scientific justification
for mobile phone users to limit
their exposure?*

_____

*Do radio signals from mobile
phones present a risk to children?*

_____

*Is industry doing enough to protect
children when they are online,
and what is the role of parents
and teachers?*

_____

*Where can people turn
to find the latest research
and recommendations?*

_____

Industry Position

**Governments should adopt the
international limit for SAR recommended
by the WHO and require compliance
declarations from device makers based
on international technical standards**.

We encourage governments to provide
information and voluntary practical
guidance to consumers and parents,
based on the position of the WHO.

The GSMA believes parents should have
access to accurate information so they
can make up their own mind about
when and if their children should use
wireless technologies.

Concerned individuals can choose to limit
their exposure by making shorter calls,
using text messaging or using hands-free
devices that can be kept away from the
head and body. Bluetooth earpieces use
very low radio power and reduce exposure.

The SAR is determined by the highest
certified power level in laboratory
conditions. However, the actual SAR level
of the phone while operating can be well
below this value. Differing SAR values do
not mean differing levels of safety.

Resources:
World Health Organization International EMF Project
International Agency for Research on Cancer Monograph on Radiofrequency Fields
GSMA: Mobile and Health — independent expert reviews
Mobile Manufacturers Forum SAR Tick Programme
ITU EMF Guide
Article: Health Council of the Netherlands finds it "highly unlikely" that RF EMF causes cancer

Deeper Dive

Health Authorities on the Science

A large number of studies have been performed over the last two decades to assess whether mobile phones pose a potential health risk. To date, no adverse health effects have been established as being caused by mobile phone use.

— WHO Fact Sheet 193, October 2014

RF research is continuing in a number of areas, but data currently available provides no clear or persuasive evidence of any other effects. For this reason, the Committee and the Ministry of Health continue to support the use of exposure limits for RF fields set in the current New Zealand Standard, which is based on guidelines published by an international scientific body recognised by the WHO for its independence and expertise in this area. Those guidelines were first published in 1998 and endorsed, following a review of more recent research, in 2009.

— Ministry of Health (New Zealand), 2015

However, in previous reports the Scientific Council of SSM has concluded that studies of brain tumours and other tumours of the head (vestibular schwannoma, salivary gland), together with national cancer incidence statistics from different countries, are not convincing in linking mobile phone use to the occurrence of glioma or other tumours of the head region among adults. Recent studies described in this report do not change this conclusion although these have covered longer exposure periods. Scientific uncertainty remains for regular mobile phone use for time periods longer than 15 years. It is also too early to draw firm conclusions regarding risk of brain tumours in children and adolescents, but the available literature to date does not indicate an increased risk.

— Swedish Radiation Safety Authority, 2015

Overall, the epidemiological studies on mobile phone RF EMF exposure do not show an increased risk of brain tumours. Furthermore, they do not indicate an increased risk for other cancers of the head and neck region. Some studies raised questions regarding an increased risk of glioma and acoustic neuroma in heavy users of mobile phones. The results of cohort and incidence time trend studies do not support an increased risk for glioma while the possibility of an association with acoustic neuroma remains open. Epidemiological studies do not indicate increased risk for other malignant diseases, including childhood cancer.

— European Commission scientific advisory committee (SCENIHR), 2015

Personal Control Over Exposure

Mobile phone users who remain concerned about the possible risks of EMF can make small changes to reduce their exposure significantly. Mobile phones increase their transmission power when the signal is weak, when they are in motion and when they are in rural areas. To decrease exposure, callers may choose to use their mobile phone more when they are outside, in one spot and in urban areas.

Using one's mobile while		
Outdoors	Stationary	In town
generates exposure levels up to		
80% lower	50% lower	50% lower
compared to		
Indoors	Moving	In the countryside

Source: GSMA

Electromagnetic Fields and Health

Background

Research into the safety of radio signals, which has been conducted for more than 50 years, has led to the establishment of human exposure standards including safety factors that provide protection against all established health risks.

The World Health Organization (WHO) set up the International EMF Project in 1996 to assess the health and environmental effects of exposure to electromagnetic fields (EMF) from all sources. The WHO reviews on-going research and provides recommendations for research to support health risk assessments.

The strong consensus of expert groups and public health agencies, such as the WHO, is that no health risks have been established from exposure to the low-level radio signals used for mobile communications.

The WHO and the International Telecommunication Union (ITU) recommend that governments adopt the radio-frequency exposure limits developed by the International Commission on Non-Ionizing Radiation Protection (ICNIRP).

The WHO is currently conducting a risk assessment for radio frequency signals. The results are expected in 2016, including policy recommendations for governments.

Debate

Does using a mobile phone regularly, or living near a base station, have any health implications?

Are there benefits in adopting electromagnetic field (EMF) limits for mobile networks or devices?

What EMF exposure limits should be specified for base stations?

Should there be particular restrictions to protect children, pregnant women or other potentially vulnerable groups?

Industry Position

National authorities should implement EMF-related policies based on established science, in line with international recommendations and technical standards.

Large differences between national limits and international guidelines can cause confusion and increase public anxiety. Consistency is vital, and governments should:

- Base EMF-related policy on reliable information sources, including the WHO, trusted international health authorities and expert scientists.

- Set a national policy covering the siting of masts, balancing effective network roll out with consideration of public concerns.

- Verify that mobile operators are compliant with international radio frequency levels using technical standards from organisations such as the International Electrotechnical Commission (IEC) and ITU.

- Actively communicate with the public, based on the positions of the WHO, to address concerns.

Parents should have access to accurate information so they can decide when and if their children should use mobile phones. The current WHO position is that international safety guidelines protect everyone in the population with a large safety factor, and that there is no scientific basis to restrict children's use of phones or the locations of base stations.

The mobile industry works with national and local governments to help address public concern about mobile communications. Adoption of evidence-based national policies concerning exposure limits and antenna siting, public consultations and information can reassure citizens.

Ongoing, high-quality research is necessary to support health-risk assessments, develop safety standards and provide information to inform policy development. Studies should follow good laboratory practice for EMF research and be governed by contracts that encourage open publication of findings in peer-reviewed scientific literature.

Resources:
World Health Organization International EMF Project
GSMA: Arbitrary Radio Frequency Exposure Limits — Impact on 4G Network Deployment
GSMA: Mobile and Health — independent expert reviews
GSMA: LTE Technology and Health
ITU-T activities on EMF
ITU EMF Guide
Article: New research shows no health risk from wireless networks in schools or at home says Swedish safety authority report

Deeper Dive

A Global Look at Mobile Network Exposure Limits

The World Health Organization (WHO) endorses the guidelines of the International Commission for Non-Ionizing Radiation Protection (ICNIRP) and encourages countries to adopt them. While many countries have adopted this recommendation, some have adopted other limits or additional measures regarding the siting of base stations.

This map shows the approach to radio frequency (RF) exposure limits countries have adopted for mobile communication antenna sites. Much of the world follows the ICNIRP 1998 guidelines or those of the US Federal Communications

Effective RF limit

● ICNIRP 1998

● FFC

● Other

● Unknown

Commission. In some cases (e.g., China and Russia) historical limits have not been updated to reflect more recent scientific knowledge. In other cases, RF limits applicable to mobile networks may be the result of arbitrary reductions, as a political response to public concern.

Excluding countries or territories with unknown limits, 124 apply ICNIRP, 11 follow the FCC limits from 1996, and 37 have other limits. The map uses only one colour for the 'other' category, however, there are many differences between these countries in the limit values and their application.

eWaste

Background

Electronic waste — also known as e-waste or waste electrical and electronic equipment (WEEE) — is a type of waste generated when devices related to the Information and Communications Technology (ICT) industry reach the end of their life. Parts and materials that make up e-waste usually contain precious or high-value metals that can be recycled at the end of a device's useful life. However, they can also contain hazardous materials that must be treated responsibly and in compliance with environmental legislation.

As part of the ICT sector, mobile operators generate e-waste during periods of technological renewal and also through the normal supply of products to customers (such as routers, mobile phones and tablets).

Mobile operators around the world have developed WEEE management programmes both as compliance measures to conform to current legislation, and also due to their desire to meet their own sustainability and corporate social responsibility goals.

However, in some regions, such as Latin American, there is a lack of legal frameworks specifically covering e-waste management. Unfortunately, this also means there is a lack of clarity around the concept of extended producer responsibility (EPR).

Usually EPR rules firmly establish the roles and responsibilities of producers, importers and distributors for equipment in the e-waste chain. The absence of clear rules means operators in Latin America are finding it difficult to manage the e-waste

generated through their operations. In some cases, they have even had to take on 100 per cent of the operational and financial responsibility for the management of their customers' e-waste, whereas in most other regions the responsibility is shared among a range of parties including equipment manufacturers, importers and distributors.

In addition, operators have faced other challenges such as a dearth of qualified e-waste managers in some countries, the high costs of e-waste transport and storage, and restrictions (due to the Basel Convention) on the export of equipment to countries where it could be treated appropriately.

Debate

How should the responsibility for processing e-waste be shared out among a range of industry parties including operators, equipment manufacturers, importers and distributors?

Industry Position

The effective management of WEEE at a country and company level must be based on specific regulatory frameworks that recognise the environmental risks that e-waste presents. This is to ensure there is no ambiguity among the various parties who are responsible for e-waste management as to how they must act in order to conform to the agreed guidelines.

Mobile operators have long recognised the importance of WEEE management. This is why, in regions such as Latin America, they have actively sought to draw attention to loopholes in the legal system and communicate the challenges they have faced during the development of their WEEE management programmes. Moreover, they continue to look for ways to collaborate with the environmental authorities in order to define effective legal frameworks that promote environmentally responsible WEEE management.

With this in mind, they have come up with a number of proposals for regions where there is currently a lack of robust legal frameworks in place:

- Environmental and telecommunications authorities should work together to design, promote and implement policies, standards, laws, regulations and programmes for responsible WEEE management.

- Guidelines should be created by relevant environmental authorities and developed into legal frameworks for e-waste management that recognise the principle of extended producer responsibility (EPR).

- WEEE management programmes should include measures to promote recycling in order to extend the lifespan of devices. These need to explain the importance of these processes for the reuse of materials, so they can in turn increase the economic value of devices collected for re-use or recycling.

- Governments, manufacturers, importers, distributors and WEEE management companies should work together to create e-waste awareness campaigns aimed at the general public. These campaigns will help create a culture of WEEE recycling, foster buy-in across all sectors of society and drive improved results when all the parties involved begin implementing WEEE management campaigns.

Resources:
GSMA Report eWaste in Latin America

Government Access

Background

Mobile network operators are often subject to a range of laws and/or licence conditions that require them to support law enforcement and security activities in countries where they operate. These requirements vary from country to country and have an impact on the privacy of mobile customers.

Where they exist, such laws and licence conditions typically require operators to retain data about their customers' mobile service use and disclose it, including customers' personal data, to law enforcement and national security agencies on lawful demand. They may also require operators to have the ability to intercept customer communications following lawful demand.

Such laws provide a framework for the operation of law enforcement and security service surveillance and guide mobile operators in their mandatory liaison with these services.

However, in some countries, there is a lack of clarity in the legal framework to regulate the disclosure of data or lawful interception of customer communications.

This creates challenges for industry in protecting the privacy of its customers' information and their communications.

Legislation often lags behind technological developments. For example, it may be the case that obligations apply only to established telecommunications operators but not to more recent market entrants, such as those providing internet-based services, including Voice over-IP (VoIP) services, video or instant messaging services.

In response to public debate concerning the extent of government access to mobile subscriber data, a number of major telecommunications providers (such as AT&T, Deutsche Telekom, Orange, Rogers, SaskTel, Sprint, T-Mobile, TekSavvy, TeliaSonera, Telstra, Telus, Verizon, Vodafone, Wind Mobile) as well as internet companies (such as Apple, Dropbox, Facebook, Google, Pinterest, Twitter, Amazon, LinkedIn, Microsoft, Snapchat, Tumblr, Twitter, Yahoo!) publish 'transparency reports' which provide statistics relating to government requests for disclosure of such data.

Debate

What is the right legal framework to achieve a balance between governments' obligation to ensure law-enforcement and security agencies can protect citizens and the rights of citizens to privacy?

Should all providers of communication services be subject to the same interception, retention and disclosure laws on a technology-neutral basis?

Would further transparency about the number and nature of the requests that governments make of communications providers assist the debate, improve government accountability and bolster consumer confidence?

Industry Position

Governments should ensure they have a proportionate legal framework that clearly specifies the surveillance powers available to national law enforcement and security agencies.

Any interference with the right to privacy of telecommunications customers must be in accordance with the law.

The retention and disclosure of data and the interception of communications for law enforcement or security purposes should take place only under a clear legal framework and using the proper process and authorisation specified by that framework.

There should be a legal process available to telecommunications providers to challenge requests which they believe to be outside the scope of the relevant laws.

The framework should be transparent, proportionate, justified and compatible with human rights principles, including obligations under applicable international human rights conventions, such as the International Convention on Civil and Political Rights.

Given the expanding range of communications services, the legal framework should be technology neutral.

Governments should provide appropriate limitations of liability or indemnify telecommunications providers against legal claims brought in respect of compliance with requests and obligations for the retention, disclosure and interception of communications and data.

The costs of complying with all laws covering the interception of communications, and the retention and disclosure of data should be borne by governments. Such costs and the basis for their calculation should be agreed in advance.

The GSMA and its members are supportive of initiatives that seek to increase government transparency and the publication by government of statistics related to requests for access to customer data.

Resources:
Guiding Principles on Business and Human Rights: Implementing the United Nations "Protect, Respect and Remedy" Framework
Malone v. The United Kingdom, Application No. 8691/79, Judgement of 2 August 1984 of the ECJ
High Court Judgement ruling unlawful the UK government 'Data Retention and Investigatory Powers Act 2014 ("DRIPA")'
A Question of Trust – Report of the [UK] Investigatory Powers Review
Office of the Privacy Commissioner of Canada: Transparency Reporting by Private Sector Companies – Comparative Analysis

National Regulatory Approaches to Government Access

Increasingly, as in the UK, France, Germany and Australia, laws are being proposed that would require service providers to capture and retain communications data and grant the government systematic access to this information.

In the UK, communications service providers are required to separately retain a range of account and communications data and must ensure the data can be disclosed in a timely manner to UK law enforcement agencies, the security services and a number of prescribed public authorities under the UK Regulation of Investigatory Powers Act (RIPA). Prescribed authorities can also seek a warrant from the Secretary of State to intercept communications.

The two main objectives of RIPA are to regulate the investigatory powers of the state and to set the legitimate expectations for citizens' privacy. As RIPA is subject to oversight by the Surveillance Commissioner and the Interception Commissioner, citizens can seek redress for alleged unlawful access to their data or communications, and service providers operating in the UK can raise concerns about the validity of requests.

In April 2014 the European Court of Justice ruled that the EU Data Retention Directive is 'invalid' as it violated two basic rights — respect for private life and protection of personal data. The European Commission has emphasised that the decision of whether or not to introduce national data retention laws is a national decision and consequently, the UK and a number of other countries in the European Union are reviewing their data retention laws, which required communications service providers to store communications data for up to two years.

Meanwhile, in May 2015, the German government has outlined plans for a new data retention law which would require telecoms companies to retain 'traffic data' relevant to communications and hand them over (under certain conditions) to Germany's law enforcement and security agencies. Germany's privacy campaigners questioned whether the plans were constitutional adding that, in their opinion, the German government had not sufficiently outlined why the retention of the data is necessary.

In July 2015, the French Parliament approved a bill that allows intelligence agencies to tap phones and emails without seeking permission from a judge. The new law requires communications providers and Internet service providers to hand over customers' data upon request, if the relevant customers are linked to a 'terrorist' inquiry. Protesters from civil liberties groups claimed the bill would legalise intrusive surveillance methods without guarantees for individual freedom and privacy.

Australia's new Telecommunications (Interception and Access) Amendment (Data Retention) Act 2015 requires telecommunication service providers to retain for two years certain telecommunications metadata prescribed by regulations. This two-year retention period equals the maximum allowed under the EU's earlier Data Retention Directive which the EU's Court of Justice ruled as invalid.

Trending Towards Transparency

Many of the largest communications and internet content providers — including AT&T, Deutsche Telekom, Telenor, Verizon, Vodafone, Apple, Dropbox, Facebook, Google, LinkedIn, Microsoft, Twitter and Yahoo! — publish periodic reports showing the types and/or volume of requests from governments for user information. Typically, these 'transparency reports' include how many of these requests resulted in the disclosure of customer information. These reports reveal not only the frequency of such requests, but some detail about the kind of information accessed — customer account information; metadata, which can reveal an individual's location, interests or relationships; and the interception of communications. Although mobile operators often have no option but to comply with such requests, they are increasingly pressing for greater transparency about the nature and scale of government access.

At a time of growing public awareness and debate over government surveillance and privacy in many countries, this trend towards reporting the demands of governments for communications data (where it is legal to do so) has revealed the degree to which government intelligence and law enforcement agencies rely on such information.

The political debate is heated on both sides — those who argue that law enforcement agencies require broad access in order to fight crime, and those who rail against perceived overzealous snooping and strive to maintain citizens' right to privacy in the digital age.

Like the Internet content providers, mobile network operators may find themselves in a difficult position — bound to meet their obligations to provide lawful access while assuring their customers that they protect private user information. Transparency reporting brings valid information to the public and policymakers, raising key questions about the balance between government access and privacy.

Illegal Content

Background

Today, mobile networks not only offer traditional voice and messaging services, but also provide access to virtually all forms of digital content via the internet. In this respect, mobile operators offer the same service as any other internet service provider (ISP). This means mobile networks are inevitably used, by some, to access illegal content, ranging from pirated material that infringes intellectual property rights (IPR) to racist content or images of child sexual abuse (child pornography).

Laws regarding illegal content vary considerably. Some content, such as images of child sexual abuse, are considered illegal around the world, while other content, such as dialogue that calls for political reform, is illegal in some countries while protected by 'freedom of speech' rights in others.

Communications service providers, including mobile network operators and ISPs, are not usually liable for illegal content on their networks and services, provided they are not aware of its presence and follow certain rules e.g., 'notice and take down' processes to remove or disable access to the illegal content as soon as they are notified of its existence by the appropriate legal authority.

Mobile operators are typically alerted to illegal content by national hotline organisations or law enforcement agencies. When content is reported, operators follow procedures according to the relevant data protection, privacy and disclosure legislation. In the case of child sexual abuse content, mobile operators use terms and conditions, notice and take down processes and reporting mechanisms to keep their services free of this content.

Debate

Should all types of illegal content — from IPR infringements to child sexual abuse content — be subject to the same reporting and removal processes?

What responsibilities should fall to governments, law enforcement or industry in the policing and removal of illegal content?

Should access to illegal content on the internet be blocked by ISPs and mobile operators?

INTERPOL is pleased to support the Mobile Alliance Against Child Sexual Abuse Content which sends a clear message from its members — that there is zero tolerance of child exploitation on their network. Alliances such as this, and its willingness to work with other stakeholders and society in general, are hugely important and will serve as an example of best practice.

— Mick Moran, Assistant Director Human Trafficking and Child Exploitation. INTERPOL

Industry Position

The mobile industry is committed to working with law enforcement agencies and appropriate authorities, and to having robust processes in place that enable the swift removal or disabling of confirmed instances of illegal content hosted on their services.

ISPs, including mobile operators, are not qualified to decide what is and is not illegal content, the scope of which is wide and varies between countries. As such, they should not be expected to monitor and judge third-party material, whether it is hosted on, or accessed through, their own network.

National governments decide what constitutes illegal content in their country; they should be open and transparent about which content is illegal before handing enforcement responsibility to hotlines, law enforcement agencies and industry.

The mobile industry condemns the misuse of its services for sharing child sexual abuse content. The GSMA's Mobile Alliance Against Child Sexual Abuse Content provides leadership in this area and works proactively to combat the misuse of mobile networks and services by criminals seeking to access or share child sexual abuse content.

Regarding copyright infringement and piracy, the mobile industry recognises the importance of proper compensation for rights holders and prevention of unauthorised distribution.

Resources:
GSMA Report: Hotlines — Responding to reports of illegal online content
Mobile Alliance Against Child Sexual Abuse Content
INHOPE website

Mobile Alliance Against Child Sexual Abuse Content

The Mobile Alliance Against Child Sexual Abuse Content was founded by an international group of mobile operators within the GSMA to work collectively on obstructing the use of the mobile environment by individuals or organisations wishing to consume or profit from child sexual abuse content.

Alliance members have made the commitment to:

- Implement technical mechanisms to restrict access to URLs identified by an appropriate, internationally recognised agency as hosting child sexual abuse content.

- Implement 'notice and take down' processes to enable the removal of any child sexual abuse content posted on their own services.

- Support and promote hotlines or other mechanisms for customers to report child sexual abuse content discovered on the internet or on mobile content services.

Through a combination of technical measures, co-operation and information sharing, the Mobile Alliance is working to stem, and ultimately reverse, the growth of online child sexual abuse content around the world.

The Mobile Alliance also contributes to wider efforts to eradicate online child sexual abuse content by publishing guidance and toolkits for the benefit of the whole mobile industry. For example, it has produced a guide to establishing and managing a hotline in collaboration with INHOPE, the umbrella organisation for hotlines. It also collaborates with the European Financial Coalition and the Financial Coalition Against Child Pornography.

Mobile Alliance Against Child Sexual Abuse Content

A report of suspected illegal child sexual abuse content is made by an internet user, directly or through their internet service provider (ISP) or mobile operator

National hotline or law enforcement agency (LEA) assesses the content

Illegal

Not illegal

Traced To Host Country

No Further Action

If the content is hosted in the same country as the hotline or LEA, notice and take down processes are instigated and the content is removed.

If the content is hosted in a different country, the report is passed on to INHOPE or the relevant LEA.

Some countries also add the URL to a 'block list' that allows ISPs and mobile operators to prevent access.

Internet Governance

Background

Internet governance involves a wide-array of activities related to the policy and procedures of the management of the internet. It encompasses legal and regulatory issues such as privacy, cybercrime, intellectual property rights and spam. It also is concerned with technical issues related to network management and standards, for example, and economic issues such as taxation and internet interconnection arrangements.

Because mobile industry growth is tied to the evolution of internet-enabled services and devices, decisions about the use, management and regulation of the internet will affect mobile service providers and other industry players and their customers.

Internet governance requires the inputs of diverse stakeholders, relating to their interests and expertise in technical engineering, resource management, standards and policy issues, among others. Interested and relevant stakeholders will vary from issue to issue.

Debate

Who 'owns' the internet?

Should certain countries or organisations be allowed to have greater decision-making powers than others?

How should a multi-stakeholder model be applied to internet governance?

Global Internet governance must be transparent and inclusive, ensuring full participation of governments, civil society, private sector and international organisations, so that the potential of the Internet as a powerful tool for economic and social development can be fulfilled.

— Joint press release from the governments of the USA and Brazil, June 2015

Industry Position

The multi-stakeholder model for internet governance and decision-making should be preserved and allowed to evolve.

Internet governance should not be managed through a single institution or mechanism, but be able to address a wide range of issues and challenges relevant to different stakeholders more flexibly than traditional government and intergovernmental mechanisms.

The internet should be secure, stable, trustworthy and interoperable, and no single institution or organisation can or should manage it.

Globalisation of key internet functions should be promoted — in a transparent way — to preserve the resiliency, security and stability of the internet.

Collaborative, diverse and inclusive models of internet governance decision-making are requisite to participation by the appropriate stakeholders.

The decentralised development of the internet should continue, without being controlled by any particular business model or regulatory approach.

Some questions warrant a different approach at the local, national, regional or global level. An effective and efficient multi-stakeholder model ensures that the stakeholders, within their respective roles, can participate in the consensus-building process for any specific issue.

Technical aspects related to the management and development of internet networks and architecture should be addressed through standards bodies, the Internet Engineering Task Force (IETF) and the Internet Architecture Board (IAB) and other fora.

Economic and transactional issues such as internet interconnection charges are best left to commercial negotiation, consistent with commercial law and regulatory regimes.

Resources:
Internet Society: Internet Governance
OECD Resources on Internet Governance
Centre for International Governance Innovation
Internet Governance Forum

Deeper Dive

Key Players in Internet Governance

Primary Organisations

United Nations Bodies

UN General Assembly (UNGA)

UN top-level body. Will review WSIS implementation in 2015.

World Summit on the Information Society (WSIS)

WSIS 2005 established IGF and WGEC. WSIS Action Lines C1 and C11 also relate directly to internet governance policy.

Internet Governance Forum (IGF)

UN Commission on Science & Technology (CSTD)

Working Group on Enhanced Cooperation (WGEC)

ITU

UN agency for information and communication technologies has remit for some technical standards.

Addressing Resources

Internet Corporation for Assigned Names and Numbers (ICANN)

Number Resource Organisation (NRO)

Collective body for the Regional Internet Registries (RIRs). RIRs manage the allocation registration of Internet number resources.

Architecture and Standards Development

Internet Society (ISOC)

Internet standards development, education and advocacy

World Wide Web Consortium (W3C)

Recommendations for implementation of web technologies

Other Intergovernmental Organisations

Security Policy Focus

Organisation of American States

Has adopted Inter-American Comprehensive Strategy for Cybersecurity

Shanghai Cooperative Organisation (SCO)

China, Kazakhstan, Kyrgyzstan, Russia, Tajikistan, Uzbekistan; focus on security

Council of Europe

2001 Convention (Treaty) on Cybercrime ratified by multiple countries (including non-European)

NATO

Has a policy and associated Action Plan on cyberdefence

APEC

Strategic 2010-15 goal re. security in IT infrastructure

World Trade Organisation (WTO)

Currently addressing IPR theft online and cyber-espionage

Generic Policy Focus

OECD

Published 'Principles on Internet Policy-Making' in 2012; is reviewing 2002 Security Guidelines

---

RIPE
RIR for Europe

LACNIC
RIR for LatAm and Caribbean

APNIC
RIR for Asia Pacific

ARIN
RIR for America

AfriNIC
RIR for Africa

Internet Engineering Task Force (IETF)

ISOC task force; principal global body developing (voluntary) Internet technical standards

Internet Architecture Board (IAB)

ISOC committee; focuses on long-range planning of technical/ engineering development

Internet Engineering Steering Group (IESG)

Responsible for technical management of IETF activities and the Internet standards process

Mandatory Registration of Prepaid SIMs

Background

In many countries, pay-monthly or post-paid mobile phone contracts are common. These require customers to provide proof of identification and evidence of sufficient funds before they enter into a billing arrangement with their mobile network operator.

In the case of prepaid or pay-as-you-go services, customers must purchase credit to activate their subscriber identity module (SIM) card. This can be done anonymously, as registration is not typically required.

An increasing number of governments, however, have recently introduced mandatory registration of prepaid SIM card users, primarily as a tool to counter terrorism and improve law enforcement.

The take-up of mobile identity, mobile-commerce and e-government services can be boosted by the registration of all SIM card users, as they would be able to verify their identity and log in to such services using their mobile device. Nevertheless, mandatory registration often leads to implementation challenges and unforeseen consequences, particularly in developing countries, where the majority of mobile users have prepaid SIM cards.

These challenges include:

- Failure by some mobile users to understand that their SIM cards could be deactivated, sometimes without warning, if they do not register by a certain deadline.

- Barriers that prevent some mobile users from physically registering, e.g., the distance to a registration centre.

- Limitations to prepaid SIM card distribution channels due to the registration requirement.

- The cost of implementation, which can be significant and may impact operators' ability to invest in new, innovative services and network infrastructure, particularly in remote and rural areas.

- The emergence of a black market for fraudulently-registered or stolen SIM cards, based on the desire by some mobile users, including criminals, to remain anonymous.

- Mobile user concerns related to the access, security, use and retention of their personal data, particularly in the absence of national laws on privacy and freedom of expression.

Some governments, including those of the UK and the Czech Republic, have decided against mandating registration of prepaid SIM users, concluding that the potential loopholes and implementation challenges outweigh the merits.

Debate

To what extent do the benefits of mandatory prepaid SIM registration outweigh the costs and risks?

What factors should governments consider before mandating such a policy?

Industry Position

While registration of prepaid SIM card users could offer valuable benefits to citizens and consumers, governments should not mandate it.

To date, there is no evidence that mandatory registration of prepaid SIM card users leads to a reduction in crime.

The effectiveness of prepaid SIM user registration depends on local market conditions, for example, whether citizen access to national identity documents is widespread throughout the country and whether the government maintains robust citizen identity records.

Where prepaid SIM user registration can create value and positive outcomes for consumers, mobile operators and governments will have an incentive to offer services that encourage consumers to register voluntarily.

We urge governments that are considering such a policy to examine the local market conditions, engage with industry and conduct impact assessments before introducing regulation.

Where a decision to mandate the registration of prepaid SIM users has been made, we recommend that governments take into account global best practices and allow registration mechanisms that are flexible, proportionate and relevant to the specific market.

Resources:
GSMA White Paper: Mandatory Registration of Prepaid SIM Card Users
Academic paper: The Rise of African SIM Registration: Mobility, Identity, Surveillance & Resistance, London School of Economics, November 2012
Academic Paper: Implications of Mandatory Registration of Mobile Phone Users in Africa, Deutsches Institut für Wirtschaftsforschung, 2012
GSMA Mobile Connect
Academic Paper: Privacy Rights and Prepaid Communication Services, Simon Fraser University, March 2006
Article: Assessing the Impact of SIM Registration on Network Quality (Nigeria), July 2013
Article: Global Crackdown on Phone Anonymity, Kosmopolitica, May 2013

Best Practice

The pros and cons of prepaid SIM registration will be different for each market. Governments considering a mandatory prepaid SIM registration policy should fully investigate a number of factors, including:

- Whether there is evidence that the registration exercise would improve the reliability of data available to law enforcement agencies and contribute to crime reduction, and whether a criminal could easily obtain a SIM card — locally or abroad — to avoid registration.

- The share of population holding a valid ID document.

- Whether the government keeps an up-to-date and robust record of citizen identity documents (which consumers are required to use when registering their SIM).

- Whether any geographic, demographic or cultural characteristic would affect how easily consumers could physically register a SIM in their name (e.g., those living in remote areas or informal housing, or those who are disabled).

- The ability to make all consumers aware that their existing prepaid SIM cards may be deactivated if they fail to register them by a certain deadline.

- The impact of any data protection and privacy laws on how consumers' personal details are collected, stored and potentially shared with government agencies and third parties.

- Whether the registration exercise will impose a disproportionate burden on mobile operators.

Implementation Factors

Where a decision to mandate prepaid SIM registration has been made, governments should take into account global best practices and consider the following:

Consumer-related issues

Identity verification and registration channels (How can prepaid SIM users verify their identity, and can the various registration channels cater to all consumer groups, such as those living in remote or rural areas?).

Effective public awareness campaigns (Are consumers aware that they need to register their SIMs and understand how to do this?).

Industry-related issues

Timescales for mobile operators to implement registration processes (Are they practical and realistic?).

The use, sharing and retention of SIM users' registration details (Are data retention and disclosure requirements proportionate, and do they preserve mobile users' privacy?).

Broader regulatory compliance

Regulatory enforcement and consequences of noncompliance for mobile operators (What are the regulator's enforcement powers after the registration deadline has passed?).

Mobile Device Theft

Background

Unfortunately, there are criminals who seek to gain from the trade of stolen mobile phones, feeding a black market in handsets obtained through mugging and street crime.

Policymakers in many countries are concerned about the incidence of mobile phone theft, particularly when organised crime becomes involved in the bulk export of stolen handsets to other markets.

In 1996, the GSMA launched an initiative to block stolen mobile phones, based on a shared database of the unique identifiers of handsets reported lost or stolen. Using the International Mobile Equipment Identifier (IMEI) of mobile phones, the GSMA maintains a central list — known as the IMEI Database — of all phones reported lost or stolen by mobile network operators' customers.

The efficient blocking of stolen devices on individual network Equipment Identity Registers (EIRs) depends on the secure implementation of the IMEI on all mobile handsets. The world's leading device manufacturers have agreed to support a range of measures to strengthen IMEI security, and progress is monitored by the GSMA.

Debate

What can industry do to prevent mobile phone theft?

What are the policy implications of this rising trend?

Should regulations be imposed on mobile device registration?

To what extent can device-based antitheft features complement network blocking of stolen devices, and what capabilities should those features support?

Handset theft is a growing crime and law enforcement problem in some markets where measures have not been taken to comprehensively deal with the issue. Every stolen phone causes misery, possible violence and psychological consequences for mobile users.

— James Moran, Security Director, GSMA

Industry Position

The mobile industry has led numerous initiatives and made great strides in the global fight against mobile device theft.

Although the problem of handset theft is not of the industry's creation, the industry is part of the solution. When lost or stolen mobile phones are rendered useless, they have no value, removing all incentive for thieves.

The GSMA encourages its member operators to deploy EIRs on their networks to deny connectivity to any stolen device. Operators should connect to the GSMA's IMEI Database to ensure devices stolen from their customers can be blocked on networks that use the database. These solutions have been in place on some networks and in some countries for many years and they continue to be improved and extended.

IMEI blocking has had a positive impact in many countries, but for a truly effective anti-theft campaign, a range of measures must be put in place, only some of which are within the control of the mobile industry.

The concept of a 'kill switch' allowing mobile phone users to remotely disable their stolen device has received much attention as mobile device theft has risen. The GSMA supports device-based anti-theft features and has defined feature requirements that could lead to a global solution for owners to locate or disable their lost or stolen device and to protect and deny access to personal data stored on the device. These high-level requirements have set a benchmark for anti-theft functionality while allowing the industry to innovate.

National authorities have a significant role to play in combatting this criminal activity. It is critical that they engage constructively with the industry to ensure the distribution of mobile devices through unauthorised channels is monitored and that action is taken against those involved in the theft or distribution of stolen devices.

A coherent regional information-sharing approach involving all relevant stakeholders would make national measures more effective.

Some national authorities have proposed national 'whitelists' to combat mobile terminal theft. The GSMA opposes this approach, which could impede the free movement of mobile devices around the world and would be considered illegal in some countries.

Resources:
OAS Briefing Paper on the Theft of Mobile Terminal Equipment
IMEI Database
Security Principles Related to Handset Theft
Anti-Theft Device Feature Requirements
IMEI Security Weakness Reporting and Correction Process
Case Study: Mobile Phone Theft in Costa Rica
Q&A: Consumer Precautions Against Mobile Phone Theft

Deeper Dive

Safeguards in Mobile Handset Manufacturing

Since 1996, the GSMA has promoted the use of Equipment Identity Registers (EIRs) among mobile network operators to ensure stolen handsets can be barred from networks by using the handsets' IMEI numbers. EIR effectiveness, however, is largely dependent on a secure implementation of the IMEI, and EIR deployment should be complemented by the efforts of the handset manufacturing community to ensure all handsets delivered to market incorporate appropriate security features. The following security principles help handset manufacturers protect the platform on which the IMEI mechanism is stored.

Principle 1
Implement safeguards for uploading, downloading and storing executable code and sensitive data related to the IMEI implementation.

Principle 2
Protect components' executable code and sensitive data related to the IMEI implementation.

Principle 3
Protect against exchange of data and software between devices.

Principle 4
Protect IMEI executable code and sensitive data from external attacks.

Principle 5
Prevent the download of previous software versions.

Principle 6
Detect and respond to unauthorised tampering.

Principle 7
Apply software quality measures for all sensitive functions.

Principle 8
Prevent hidden areas from accessing or modifying executable code or sensitive data related to IMEI implementation.

Principle 9
Prevent the substitution of hardware components.

Mobile Security

Background

Security attacks threaten all forms of ICT, including mobile technologies. Consumer devices such as mobile handsets are targeted for a variety of reasons, from changing the IMEI number of a mobile phone to re-enable it after theft, through to data extraction or the use of malware to perform functions that have the potential to cause harm to users.

Mobile networks use encryption technologies to make it difficult for criminals to eavesdrop on calls or to intercept data traffic. Legal barriers to the deployment of cryptographic technologies have been reduced in recent years and this has allowed mobile technologies to incorporate stronger and better algorithms and protocols, which remain of significant interest to hackers and security researchers.

The emerging area of Near Field Communications (NFC) has raised the concept of electronic pickpocketing, or hacking into someone's NFC-enabled account from close proximity. This potential threat continues to receive more attention as NFC applications gain market traction and the role of the SIM as a secure platform for the hosting and execution of sensitive services becomes key.

The GSMA plays a key role in coordinating the industry response to security incidents, and it cooperates with a range of stakeholders including its operator members, device manufacturers and infrastructure suppliers to ensure a timely and appropriate response to threats that could affect services, networks or devices.

Debate

How secure are mobile voice and data technologies?

How significant is the threat of mobile malware, and what is being done to mitigate the risks?

Do emerging technologies and services create new opportunities for criminals to steal information, access user accounts or otherwise compromise the security and safety of mobile networks and those that use them?

Industry Position

The protection and privacy of customer communications is at the forefront of operators' concerns.

The mobile industry makes every reasonable effort to protect the privacy and integrity of customer and network communications. The barriers to compromising mobile security are very high and research into possible vulnerabilities has generally been of an academic nature.

While no security technology is guaranteed to be unbreakable, practical attacks on GSM-based services are extremely rare, as they would require considerable resources, including specialised equipment, computer processing power and a high level of technical expertise beyond the capability of most people.

Reports of GSM eavesdropping are not uncommon, but such attacks have not taken place on a wide scale, and UMTS and LTE networks are considerably better protected against eavesdropping risks.

Although mobile malware has not reached predicted epidemic levels, the GSMA is aware of the potential risks and its Mobile Malware Group coordinates the operator response to identified threats. The group facilitates the prompt exchange of information between industry stakeholders and encourages best practice to manage and handle malware by producing comprehensive guidelines for its members.

The GSMA supports global security standards for emerging services and acknowledges the role that SIM-based secure elements can play, as an alternative to embedding the security into the handset or an external digital card (microSD), because the SIM card has proven itself to be resilient to attack.

The GSMA constantly monitors the activities of hacker groups, as well as researchers, innovators and a range of industry stakeholders to improve the security of communications networks. Our ability to learn and adapt can be seen from the security improvements from one generation of mobile technology to the next.

Resources:
GSMA Statement on Media Reports Relating to the Breaking of GSM Encryption
GSMA Security Accreditation Scheme
GSMA Security Advice for Mobile Phone Users

Industry Vigilance to Protect Mobile Customers

The GSMA manages numerous working groups composed of subject-matter experts from GSMA member companies. Each working group focuses on an issue that requires cross-industry cooperation, and mobile security is one of these. The GSMA Fraud and Security Group is responsible for technical security matters, maintenance and development of security algorithms, refinement of technical solutions to combat fraud and dissemination of security warnings and advice to GSMA members.

Fraud and Security Group Activities

• Identify and analyse security risks to which network operators are exposed.

• Advise network operators of the latest best practice being adopted in terms of technical security.

• Submit operator requirements to international standards bodies.

• Advise on technical solutions to combat fraud.

• Maintain and enhance mobile security levels.

• Meet changing threats.

With its wide remit and the ever-changing nature of security in information and communication technology (ICT), GSMA is highly responsive to security events and new potential risks. For example, in 2013 when security researcher Karsten Nohl alerted the GSMA to a potential weakness in SIM encryption, the GSMA was able to assess the risks, issue a range of briefings to its members and provide guidance on the countermeasures operators could take. In that instance, only a minority of SIMs produced against older standards were found to be vulnerable. The swift and comprehensive response, led by the GSMA, was widely recognised and commended.

Fraud and Security Group Subgroups

Mobile Malware Group
Responsible for coordinating the operator response to emerging threats posed by mobile malware and mobile device vulnerabilities.

Device Security Group
Responsible for device-related security threats that capture the attention and concern of regulators, the media and concerned users.

Fraud and Security Architecture Group
Focused on technical fraud and security risks and assessing implications of the introduction of new mobile technologies, architectures and network elements to ensure security is designed in from the outset by liaising with relevant standards development organisations.

Fraud and Security Communications Group
Facilitates the gathering, analysis and dissemination of information to assist GSMA members better detect, manage, and prevent fraud.

Roaming and Interconnect Fraud and Security Group
Responsible for raising awareness of signalling protocol risks and to reduce the potential for known weaknesses by investigating and recommending countermeasures and mitigation strategies.

Security Assurance Group
Develops, manages and promotes GSMA's supplier security certification schemes and provides a governance framework for security assurance of industry suppliers by defining certification methodologies, security requirements and associated guidelines for all participating stakeholders.

Number-Resource Misuse and Fraud

Background

Many countries have serious concerns about number-resource misuse, a practice whereby calls never reach the destination indicated by the international country code, but are terminated prematurely through carrier and/or content provider collusion to revenue-generating content services without the knowledge of the ITU-T-assigned number range holder.

This abuse puts such calls outside any national regulatory controls on premium-rate and revenue-share call arrangements, and is a key contributing factor to International Revenue Share Fraud (IRSF) perpetrated against telephone networks and their customers. Perpetrators of IRSF are motivated to generate incoming traffic to their own services with no intention of paying the originating network for the calls. They then receive payment quickly, long before other parts of the settlement.

Misuse also affects legitimate telephony traffic, through the side-effects of blocked high-risk number ranges.

Debate

How can regulators, number-range holders and other industry players collaborate to address this type of misuse and the resulting fraud?

Industry Position

Number-resource misuse has a significant economic impact for many countries, so multi-stakeholder collaboration is key.

The telecommunications fraud carried out as a consequence of number-resource misuse is one of the topics being addressed by the GSMA Fraud and Security Group, a global conduit for best practice with respect to fraud and security management for mobile network operators. The Fraud and Security Group's main focus is to drive industry management of mobile fraud and security matters to protect mobile operators and consumers, and safeguard the mobile industry's trusted reputation.

The Fraud and Security Group supports European Union guidelines under which national regulators can instruct communications providers to withhold payment to downstream traffic partners in cases of suspected fraud and misuse.

The group believes that national regulators can help communications providers reduce the risk of number-resource misuse by enforcing stricter management of national numbering resources. Specifically, regulators can:

- Ensure national numbering plans are easily available, accurate and comprehensive.

- Implement stricter controls over the assignment of national number ranges to applicants and ensure the ranges are used for the purpose for which they have been assigned.

- Implement stricter controls over leasing of number ranges by number-range assignees to third parties.

The Fraud and Security Group shares abused number ranges used for fraud among its members and with other fraud management industry bodies. It also works with leading international transit carriers to reduce the risk of fraud that arises as a result of number-resource misuse.

Resources:
ITU-T: Misuse of an E.164 International Numbering Resource
GSMAs fraud management resources are available only to members

Top 10 Countries Whose Numbering Resources Are Being Abused

Recommended Operator Controls to Reduce Exposure to Fraud from Number-Resource Misuse

Implement controls at the point of subscriber acquisition and controls to prevent account takeover.

Remove the conference or multi-call facility from a mobile connection unless specifically requested, as fraudsters can use this feature to establish up to six simultaneous calls.

Remove the ability to call forward to international destinations, particularly to countries whose numbering plans are commonly misused.

Utilise the GSMA high-risk ranges list, so that unusual call patterns to known fraudulent destinations can raise alarms or be blocked.

Ensure roaming usage reports received from other networks are monitored 24x7, preferably through an automated system.

Ensure that up-to-date tariffs, particularly for premium numbers, are applied within roaming agreements.

Implement the Barring of International Calls Except to Home Country (BOIEXH) function for new or high-risk subscriptions.

Privacy

Background

Research shows that mobile customers are concerned about their privacy and want simple and clear choices for controlling how their private information is used. They also want to know they can trust companies with their data. A lack of trust can act as a barrier to growth in economies that are increasingly data driven.

One of the major challenges faced by the growth of the mobile internet is that the security and privacy of people's personal information is regulated by a patchwork of geographically bound privacy regulations, while the mobile internet service is, by definition, international. Furthermore, in many jurisdictions the regulations governing how customer data is collected, processed and stored vary considerably between market participants. For example, rules governing how personal data is treated by mobile operators may be different to those governing how it can be used by internet players.

This misalignment between national privacy laws and global standard practices that have developed within the internet ecosystem makes it difficult for operators to provide customers with a consistent user experience. Equally, the misalignment may cause legal uncertainty for operators, which can deter investment and innovation. The inconsistent levels of protection also create risks that consumers might unwittingly provide easy access to their personal data, leaving them exposed to unwanted or undesirable outcomes such as identity theft and fraud.

Debate

How can policymakers help create a privacy framework that supports innovation in data use while balancing the need for privacy across borders, irrespective of the technology involved?

How is responsibility for ensuring privacy across borders best distributed across the mobile internet value chain?

What role does self-regulation play in a continually evolving technology environment?

What should be done to allow data to be used to support the social good and meet pressing public policy needs?

Industry Position

Currently, the wide range of services available through mobile devices offers varying degrees of privacy protection. To give customers confidence that their personal data is being properly protected, irrespective of service or device, a consistent level of protection must be provided.

Mobile operators believe that customer confidence and trust can only be fully achieved when users feel their privacy is appropriately protected.

The necessary safeguards should derive from a combination of internationally agreed approaches, national legislation and industry action. Governments should ensure legislation is technology-neutral and that its rules are applied consistently to all players in the internet ecosystem.

Because of the high level of innovation in mobile services, legislation should focus on the overall risk to an individual's privacy, rather than attempting to legislate for specific types of data. For example, legislation must deal with the risk to an individual arising from a range of different data types and contexts, rather than focusing on individual data types.

The mobile industry should ensure privacy risks are considered when designing new apps and services, and develop solutions that provide consumers with simple ways to understand their privacy choices and control their data.

The GSMA is committed to working with stakeholders from across the mobile industry to develop a consistent approach to privacy protection and promote trust in mobile services.

Resources:
GSMA: Consumer Research Insights and Considerations for Policymakers
GSMA Mobile and Privacy
Mobile Privacy Principles
Privacy Design Guidelines for Mobile Application Development

Mobile Privacy Principles

The GSMA has published a set of universal Mobile Privacy Principles that describe how mobile consumers' privacy should be respected and protected.

Openness, transparency and notice
Responsible persons (e.g., application or service providers) shall be open and honest with users and will ensure users are provided with clear, prominent and timely information regarding their identity and data privacy practices.

Purpose and use
The access, collection, sharing, disclosure and further use of users' personal information shall be limited to legitimate business purposes, such as providing applications or services as requested by users, or to otherwise meet legal obligations.

User choice and control
Users shall be given opportunities to exercise meaningful choice, and control over their personal information.

Data minimisation and retention
Only the minimum personal information necessary to meet legitimate business purposes should be collected and otherwise accessed and used. Personal information must not be kept for longer than is necessary for those legitimate business purposes or to meet legal obligations.

Respect user rights
Users should be provided with information about, and an easy means to exercise, their rights over the use of their personal information.

Security
Personal information must be protected, using reasonable safeguards appropriate to the sensitivity of the information.

Education

Users should be provided with information about privacy and security issues and ways to manage and protect their privacy.

Children and adolescents

An application or service that is directed at children and adolescents should ensure that the collection, access and use of personal information is appropriate in all given circumstances and compatible with national law.

Key areas of concern for privacy of mobile data

Data Capture	Data Security	Data Usage
What is my data used for? Is it used for commercial gain? For advertisements? Do I have a say in that?	Is my data safe? How is it being protected? What do I do if it gets compromised?	What happens to my personal data when I use my mobile? What data is collected? Who uses the data? For how long it is retained?
83% of respondents feel 3rd parties should seek permission before using their personal data	**88%** of respondents feel safe-guarding personal information is very important	**72%** of respondents are concerned about sharing the exact location of their mobile

Source: Futuresight, GSMA – User Perspectives on Mobile Privacy (2012)

Signal Inhibitors

Background

Signal inhibitors, also known as Jammers, are devices that generate interference in order to intentionally disrupt communication services. In the case of mobile services, they interfere with the communication between the mobile terminal and the base station.

In some regions, such as Latin America, signal inhibitors are used to prevent the illegitimate use of mobile phones in sensitive areas, such as prisons. However, blocking the signal does not address the root cause of the problem — wireless devices illegally ending up in the hands of inmates who then use them for illegitimate purposes.

Mobile network operators invest heavily to provide coverage and capacity through the installation of radio base stations. However, the indiscriminate use of signal inhibitors compromises these investments by causing extensive disruption to the operation of mobile networks, reducing coverage and leading to the deterioration of service for consumers.

Debate

Should governments or private organisations be allowed to use signal inhibitors that interfere with the provision of voice and data mobile services to consumers?

Should the marketing and sale of signal inhibitors to private individuals and organisations be prohibited?

Industry Position

In some Latin American countries, such as El Salvador, Guatemala, Honduras and Colombia, governments are promoting the deployment of signal inhibitors to limit the use of mobile services in prisons. The GSMA and its members are committed to working with governments to use technology as an aid for keeping mobile phones out of sensitive areas, as well as cooperating on efforts to detect, track and prevent the use of smuggled devices.

However, it is vital that a long-term, practical solution is found that doesn't negatively impact legitimate users, nor affect the substantial investments that mobile operators have made to improve their coverage.

The nature of radio signals makes it virtually impossible to ensure that the interference generated by inhibitors is confined, for example, within the walls of a building. Consequently, the interference caused by signal inhibitors affects citizens, services and public safety. It restricts network coverage and has a negative effect on the quality of services delivered to mobile users. Furthermore, inhibitors cause problems for other critical services that rely on mobile communications. For example, during an emergency they could limit the ability of mobile users to contact emergency services via numbers such as '999' or '911', and they can interfere with the operation of mobile connected alarms or personal health devices.

The industry's position is that signal inhibitors should only be used as a last resort and only deployed in coordination with operators. This coordination must continue for the total duration of the deployment of the devices — from installation through to deactivation — to ensure that interference is minimised in adjacent areas and legitimate mobile phone users are not affected. Furthermore, to protect the public interest and safeguard the delivery of mobile services, regulatory authorities should ban the use of signal inhibitors by private entities and establish sanctions for private entities that use or commercialise them without permission from relevant authorities.

Nevertheless, strengthening security to prevent wireless devices being smuggled into sensitive areas, such as prisons, is the most effective measure against the illegal use of mobile devices in these areas, as it would not affect the rights of legitimate users of mobile services.

Resources:
GSMA public policy position on signal inhibitors in Latin America

Spam

Background

'Spam' refers to bulk unsolicited messages. Most spam is intended to defraud or scam the recipient.

Attack techniques constantly change, as spammers identify new opportunities in the ever-changing technological, social, political and economic environment. Spammers are not inclined to obey local or international laws.

Spam detection and prevention techniques must continually evolve to stay ahead of spammers. The only effective way to prevent spam is to stop the messages from being delivered.

Spam is being discussed at many international law enforcement conferences and by multi-stakeholder organisations, including the Internet Engineering Task Force and the Internet Governance Forum.

Downloadable smartphone apps have opened another avenue for spammers to propagate unwanted messages and fraudulent content.

Debate

How can spam-related threats be addressed in the context of mobile services?

Industry Position

The GSMA and its members recommend combatting mobile spam by improving industry intelligence and collaborating with local law enforcement whenever possible.

Technology allows spammers to easily cross borders and evade local laws and law enforcement. Effectively addressing the problem requires global collaboration in law enforcement and technology.

Mobile spam damages the industry by increasing operator costs and reducing consumer trust. Mobile network operators should defend against these threats and continually protect the quality of the mobile service while reinforcing subscriber trust.

The GSMA offers a Mobile Spam Code of Practice, a coordinated effort among mobile operators to prevent SMS spam on mobile networks.

The GSMA recommends public spam reporting services which enable consumers to easily report spam via universal mechanisms, such as the short code '7726', which spells 'spam' on most device keyboards. These reports help participating operators take appropriate action to terminate spam attacks and improve their spam defence tactics. National, industry-coordinated efforts are encouraged to maximise the impact of prevention activities.

We believe that an international telecoms treaty is not the correct instrument for combating spam, as this could potentially raise sensitive issues regarding commercial or political free speech.

Formal regulatory measures to address spam should be introduced as a last resort, focused at the national level and only implemented after detailed impact assessments have been conducted.

Resources:
GSMA Mobile Spam Code of Practice

Mobile Spam Code of Practice

The Mobile Spam Code of Practice has been devised to protect the secure and trusted environment of mobile services to ensure customers receive minimal amounts of spam sent via SMS and MMS. The code takes a firm stance on how to deal with mobile spam messages that are either fraudulent or unsolicited commercial messages.

Participation by mobile operators is voluntary and applies specifically to three types of unsolicited SMS and MMS messages:

Commercial messages sent to customers without their consent.

Commercial messages sent to customers encouraging them directly or indirectly to call or send a message to a premium rate number.

Bulk unlawful or fraudulent messages sent to customers (e.g., faking, spoofing or scam messages).

Under the code, the mobile operators that are signatories commit to:

Include anti-spam conditions in all new contracts with third-party suppliers.

Provide a mechanism that ensures appropriate customer consent and effective customer control with respect to mobile operators' own marketing communications.

Work co-operatively with other mobile operators, including those who are not signatories to the code.

Provide customers with information and resources to help them minimise the levels and impact of mobile spam.

Undertake other anti-spam activities, such as ensuring that an anti-spam policy is in place, prohibiting the use of the mobile network for initiating or sending mobile spam, and adopting GSMA-recommended techniques for detecting and dealing with the international transmission of fraudulent mobile spam.

Encourage governments and regulators to support this industry initiative.

GSMA Intelligence

GSMA Intelligence is an extensive and growing resource for GSMA members, associate members and other organisations interested in understanding the mobile industry. Through industry data collection and aggregation, market research and analysis, GSMA Intelligence provides a valuable view of the mobile industry around the globe.

Global coverage

GSMA Intelligence publishes data and insights spanning 237 countries, more than 1,400 mobile network operators and over 1,200 mobile virtual network operators (MVNOs). Comprising approximately 26 million individual data points, GSMA Intelligence combines historical and forecast data from the beginnings of the industry in 1979 forward to a five-year outlook. New data is added every day.

Numerous data types

The data includes metrics on mobile subscribers and connections, operational and financial data, and socio-economic measures that complement the core data sets. Primary research conducted by the GSMA adds insight into more than 3,500 network deployments and more than 450 spectrum auctions to date. White papers and reports from across the GSMA and weekly bulletins are also available as part of the service.

Powerful data tools

Information in GSMA Intelligence is made easy to use by a range of data-selection tools: multifaceted search, rankings, filters, dashboards, a real-time data and news feed, as well as the ability to export data into Excel, or graphs and charts into presentations.

https://gsmaintelligence.com
info@gsmaintelligence.com

Global Market
Source: GSMA

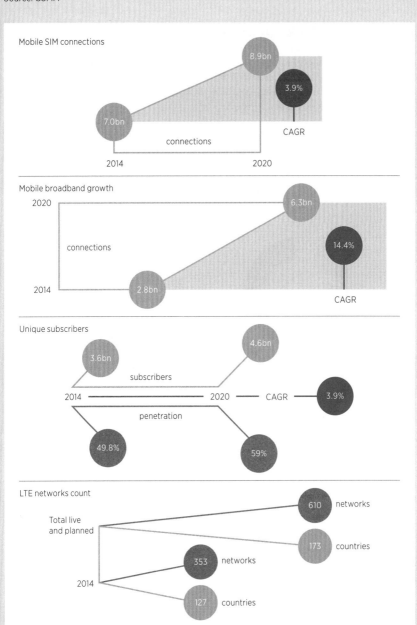

Mobile SIM connections

8,9bn
7.0bn
3.9% CAGR
connections
2014   2020

Mobile broadband growth

2020
6.3bn
connections
14.4% CAGR
2.8bn
2014

Unique subscribers

3.6bn
4.6bn
subscribers
2014   2020   CAGR   3.9%
penetration
49.8%
59%

LTE networks count

Total live
and planned
610 networks
173 countries
353 networks
2014
127 countries

CAGR: compound annual growth rate

Unique subscriber penetration by region
Source: GSMA Intelligence

The global unique subscriber base grew by 5.0 per cent during 2014: growth is forecast to continue, but at a slower rate of 3.9 per cent out to 2020. However, this growth is far from uniform across the regions of the world. Growth is now largely coming from developing markets, which are forecast to add nearly 900 million subscribers over the next six years, compared to only 50 million new additions in developed markets over the same period.

Unique subscriber penetration rates vary significantly across regions. Europe has the highest penetration rate on average, followed by North America and then the Commonwealth of Independent States. Sub-Saharan Africa had the lowest penetration rate at the end of 2014 at 38 per cent of the population, despite having seen the fastest subscriber growth of any region over the past decade.

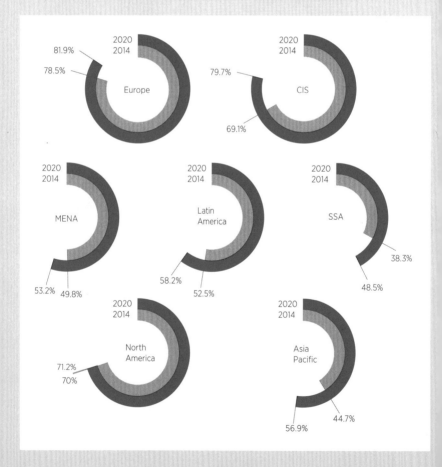

Mobile operator group global ranking by connections Q1 2015
Source: GSMA Intelligence, company reports

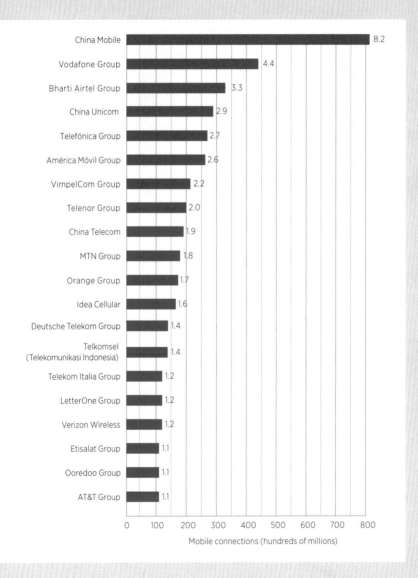

Global connection trends
Source: GSMA Intelligence

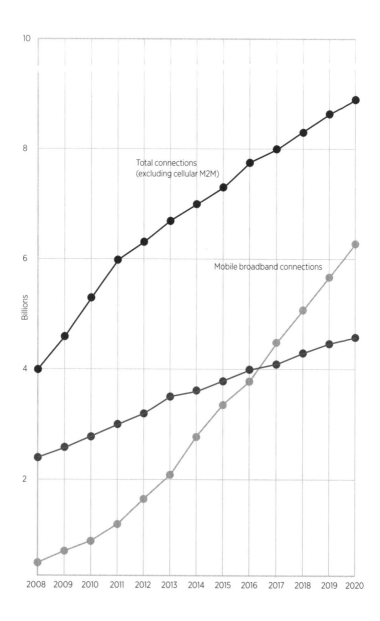

Global 4G-LTE connections forecast 2010-2020
Source: GSMA Intelligence

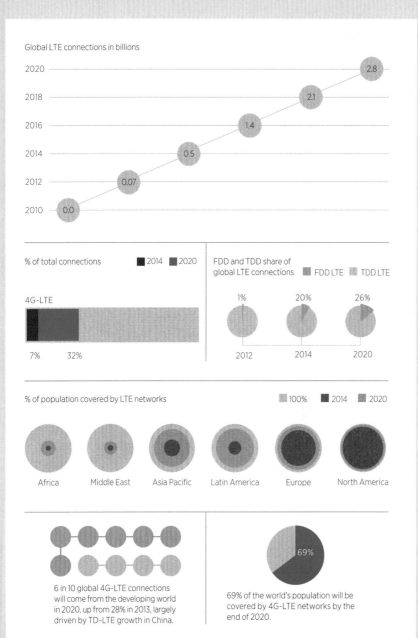

Global LTE connections in billions

2020	2.8
2018	2.1
2016	1.4
2014	0.5
2012	0.07
2010	0.0

% of total connections   ■ 2014 ■ 2020

FDD and TDD share of global LTE connections   ■ FDD LTE ■ TDD LTE

4G-LTE

7%   32%

1% (2012)   20% (2014)   26% (2020)

% of population covered by LTE networks   ■ 100% ■ 2014 ■ 2020

Africa   Middle East   Asia Pacific   Latin America   Europe   North America

6 in 10 global 4G-LTE connections will come from the developing world in 2020, up from 28% in 2013, largely driven by TD-LTE growth in China.

69%

69% of the world's population will be covered by 4G-LTE networks by the end of 2020.

Percentage of mobile operators that receive spectrum

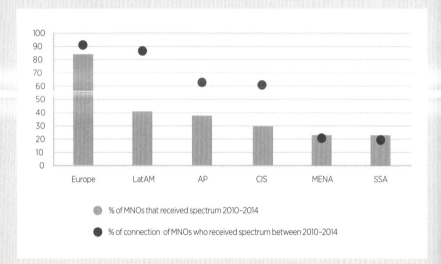

Share of LTE deployments by frequency band, by region (Jan-2015)
Source: GSMA Intelligence

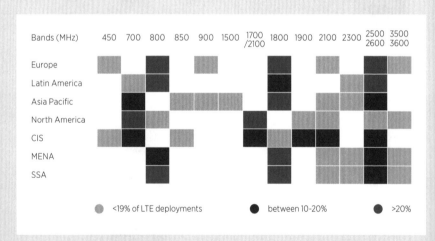

Total (direct and indirect) contribution to GDP
(2014 $bn)

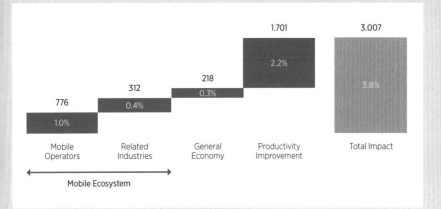

Total mobile contribution to GDP out to 2020
Value added ($bn, bars) and as a % of GDP (top)

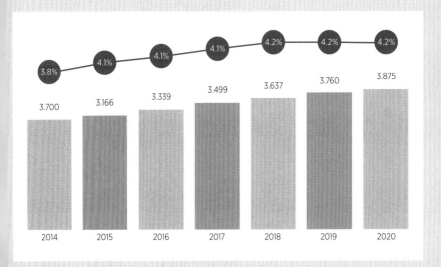

Number of Live Mobile Money Services for the unbanked by country
(December 2014)

● One Mobile Money Service

● Two Mobile Money Services

● Three or more Mobile Money
Services

● Interoperable Markets

Percentage of Investment in Mobile Money
2013 v 2014

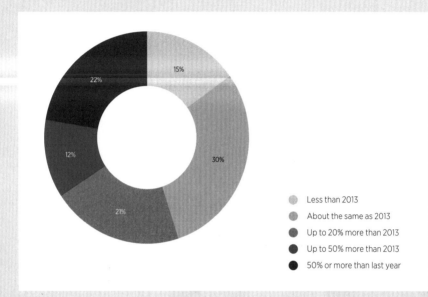

Less than 2013
About the same as 2013
Up to 20% more than 2013
Up to 50% more than 2013
50% or more than last year

Percentage of developing markets with mobile money per region
(December 2014)

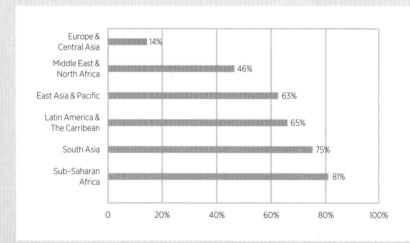